Energy Analysis: A New Public Policy Tool

AAAS Selected Symposia Series

Published by Westview Press
5500 Central Avenue, Boulder, Colorado

for the

American Association for the Advancement of Science
1776 Massachusetts Ave., N.W., Washington, D.C.

Energy Analysis: A New Public Policy Tool

Edited by Martha W. Gilliland

AAAS Selected Symposium 9

AAAS Selected Symposia Series

Published in 1978 in the United States of America by

    Westview Press, Inc.
    5500 Central Avenue
    Boulder, Colorado 80301
    Frederick A. Praeger, Publisher and Editorial Director

Library of Congress Number: 77-15895
ISBN: 0-89158-437-4

Printed and bound in the United States of America

About the Book

In meeting its energy policy responsibilities, Congress has identified a need for more adequate policymaking tools for evaluating and comparing energy alternatives. Although energy analysis has generated a great deal of interest, that interest reflects dissatisfaction with existing tools as much as it reflects the utility of energy analysis itself. The papers in this volume are intended to clarify that utility and to identify and discuss the kinds of significant policy questions that can be answered by energy analysis, either alone or in conjunction with other programs. The contributions represent diverse viewpoints on methods of energy analysis and on factors deemed directly relevant to policy areas. The introduction delineates the nature and substance of the controversy surrounding aspects of energy analysis and discusses how each of the five papers pertains to the controversy.

Contents

List of Figures

List of Tables

Foreword

The *AAAS Selected Symposia Series* was begun in 1977 to provide a means for more permanently recording and more widely disseminating some of the valuable material which is discussed at the AAAS Annual National Meetings. The volumes in this *Series* are based on symposia held at the Meetings which address topics of current and continuing significance, both within and among the sciences, and in the areas in which science and technology impact on public policy. The *Series* format is designed to provide for rapid dissemination of information, so the papers are not typeset but are reproduced directly from the camera copy submitted by the authors, without copy editing. The papers are reviewed and edited by the symposia organizers who then become the editors of the various volumes. Most papers published in this *Series* are original contributions which have not been previously published, although in some cases additional papers from other sources have been added by an editor to provide a more comprehensive view of a particular topic. Symposia may be reports of new research or reviews of established work, particularly work of an interdisciplinary nature, since the AAAS Annual Meeting typically embraces the full range of the sciences and their societal implications.

WILLIAM D. CAREY
Executive Officer
American Association for
the Advancement of Science

About the Editor

*Martha W. Gilliland is director of Energy Policy Studies,
Inc., in El Paso, Texas, and a research fellow in the Science
and Public Policy Program, University of Oklahoma, Norman.
With other members of the University of Oklahoma Technology
Assessment Group, she was coauthor of* Our Energy Future: The
Role of Research Development and Demonstration in Reaching a
National Consensus on Energy Supply *(University of Oklahoma
Press, 1977) and has published an article on energy analysis
in a recent issue of* Science *("Energy Analysis and Public
Policy," September 26, 1975).*

Introduction

Martha W. Gilliland

The aim of energy analysis is to quantify the energy flows inherent in all systems. As applied to economic systems, it is concerned with the energy flows inherent in the production of goods and services. The claims made for the policy utility of that energy flow information, however, vary across a wide spectrum. The papers included in this symposium volume represent and reflect that spectrum of claims. In setting the stage for these papers, this introduction will sketch the areas of agreement and disagreement around which the energy analysis debate revolves.

In delineating the nature and substance of the controversy over using energy analysis as a policy tool, it is convenient to distinguish between analyses at the micro level and at the macro level. This distinction applies both to the system level at which the analysis is carried out and to the system level to which results are applied. That is, analyses may be carried out for the purpose of comparing two industrial processes for producing the same commodity or for the purpose of identifying process changes which would change the energy costs of production (micro analysis) or they may be carried out for the purpose of identifying the contribution a technological system set in its environmental context makes to U. S. socioeconomic well-being (macro analysis).

There is general agreement that the more modest claims for the policy utility of micro level analyses are convincing; specifically, that a quantitative understanding of the energy flows inherent in the production of goods and services can provide a basis for better informed public policy. There is debate, however, even at this micro level about how best to measure energy flows and just what to include as an energy flow. For example, controversy is generated when efforts are made to assign energy values to labor and when second law considerations (after the second law of thermodynamics) are invoked to assign quality factors to different forms of

energy and to different tasks. There is major and intense
disagreement over the more extensive claims for the utility
of macro level analysis. The debate here flows from the
either implicit or explicit claim that energy analysis is a
better policy tool than macro economic analysis. Since each
level of analysis addresses different policy questions, full
utility and full understanding of system energetics may
require that both be carried out.

The Micro Debate

Micro analyses are generally restricted to specific or
limited numbers of processes, tasks, or technologies. That
is, they involve analyses of small systems which can be
defined precisely. When energy analyses are restricted to
specific components and seek to analyze only physical inputs
and outputs they are nearly always viewed as having utility.
This is particularly the case when the analysts present their
work as being only one of a variety of tools that may be
useful to policy makers. Who can disagree with the utility
of being able to compare the energy costs of various
processes for producing the same product or service.

Clark Bullard's paper in this symposium volume presents
an overview of the kinds of policy questions that energy
analysis has been quite effective in examining at the micro
level. In fact, as his paper indicates, the analytical
results have played a significant role in federal policy dis-
cussions on several issues. Bullard's paper represents an
overview of work carried out, beginning in 1971 (1), by the
Energy Research Group at the Center for Advanced Computation,
University of Illinois. Using an input-output model
developed at the Center, that group is now engaged in
quantifying the energy and labor impacts of many kinds of
energy production and conservation strategies (2). In industry, the term energy accounting is often used for this level
of analysis. Some industries now keep energy accounts as
well as money accounts in similar ledger fashion. Although
there remain some methodological disagreements at this level
of energy analysis, nearly everyone agrees that the results
have utility for policymaking when considered as one of a
multiplicity of factors needed for well informed
decisionmaking.*

---

*Other energy analysis research groups using somewhat
different methods for bounding the system of interest and
for measuring the energy flows include those in the Energy
Studies Unit of the University of Strathclyde, headed by
Malcolm Slesser (5) and those in the Department of Chemistry
of the University of Chicago, headed by R. Stephen Berry (6).

When the analyst assigns quantitative energy values or
weights to labor, critics surface immediately. The disagreement is not over whether labor can be viewed as a form
of energy. Obviously, to some extent, labor can be
substituted for materials or fuels. Rather, the argument is
over the energy values assigned to labor. There is no agreement on how to assign values to labor. For these reasons
some energy analysts, including those at the Center for
Advanced Computation, tabulate labor intensity separately
from the tabulation of physical inputs and outputs. Separate
tabulations of labor energy provide decisionmakers with more
information and do not blur any qualitative differences that
may exist between labor and other sources of energy.

Like the efforts to integrate values for labor into
micro energy analysis, efforts to integrate second law calculations also surface critics. At this micro level, second
law calculations refer to those given by Berg (3) and the
American Physical Society (4), where second law efficiency is
defined as the ratio of the available work which is theo-
retically required by an ideal process in order to perform a
task to the available work which is actually consumed by the
task as it is carried out with present technological systems.
This second law efficiency notion must be distinguished from
H. T. Odum's concept of second law efficiency under maximum
power (see forward in this introduction and Odum's paper in
this book), a concept that is the foundation of macro
energy analysis. It should be noted that there is little
disagreement over the theoretical ability to integrate
second law efficiency calculations of the American Physical
Society type. Rather, the disagreement is over the practical
ability and utility of these calculations. Second law
analyses require highly precise calculations based on highly
specific engineering definitions of the task for which the
energy is required. It is this specificity which is
sometimes viewed as impractical. However, it is also the
specificity in defining tasks and systems which might perform
the tasks that can lead to a more thorough understanding as
well as improvements in the manner in which energy is
consumed. Marc Ross's paper in this volume lays out some of
the characteristics and policy applications of second law
efficiency analysis.

The Macro Debate

The controversial issues associated with micro analysis
pale by comparison with those associated with macro energy
analysis. Most critics of macro analyses see it as an effort
to develop an energy theory of value. That is, it is seen
as an effort to substitute energy analysis for economic
analysis. The thrust of the comprehensive theory of energy

transformations developed by H. T. Odum is in the direction
of an alternative to macro economics. Odum's theory, which
is part of the basic science of energetics, began emerging in
1967 (7); two more recent books expand it (8). The theory
addresses both the evaluation of alternative energy strategies as well as the long range basic science goal of
understanding the energy flows in the biosphere and the place
of man-technological systems among those energy flows. It
presents a holistic view of man and nature and purports to
explain the interactions among matter, energy, and money. In
essence, the theory rests on the principle that all processes, because of physical laws, are and will be selected
for (or against) based on their contribution to maximizing
power (energy flow per unit time) in the economy. Odum, then,
views energy analysis as a tool to calculate the contribution
of existing processes to such a power maximization, to
predict it for processes under consideration, and to predict
which new processes are and will be feasible (that is, which
processes will eventually be selected for and which will be
selected against based on energy flow constraints).

In truth, proponents of macro energy analysis are not
modest in their claims for the theory. They contend that the
theory of energetics may be the needed overall framework into
which economics can fit. Over the long term, the physical
laws that govern energy flows are said to determine the value
man places on goods and services. Because it rests on
physical laws, energy analysis is not perturbed by short-term
disruptions caused by man's changing tastes or imperfect
institutions.

By comparison, economic theory is said to be inherently
dominated by short term considerations and by the fact that
it relies on an abstract model. For example, economic theory
assumes the system will seek a balance between supply and
demand for a given product. It cannot easily handle the
problems associated with the exhaustion of a finite resource. Furthermore, it omits the components of systems
which do not have money transactions associated with them,
but do have energy flows.

Economists respond by arguing that the key elements for
public policy are not the existence of physical laws, but
rather the value man puts on goods and services. They contend that energy analysis fails to integrate this absolutely
dominate consideration in its theory. In contrast, energy
analysts expect that human value choices ultimately follow
energy costs.

In fact, Odum is the first to say that economics and
market choice making mechanisms have been and are an
affective mechanism for maximizing energy flows. It is the
essential parallelism of economics and energy flows that
explains the success of economic analysis and its widespread

use. Odum's contention is, however, that enough is now
known to move to a more comprehensive and powerful theory:
macro energy analysis.

Odum's paper in this volume seeks to illustrate one
area where the use of energy analysis would lead to decisions
that run contrary to present conventional wisdom. Using
energy analysis he finds that such environmental control
technologies as cooling towers and tertiary sewage treatment
plants may cause more environmental damage than they
alleviate. This conclusion results when one calculates the
energy demands required to build and install the control
equipment. The policy question then becomes one of judging
whether the local environmental benefits are worth the
broader national environmental costs. In sum, Odum's theory
purports to give a much more accurate reading of the
relationship of man and nature than is possible with macro
economics.

In practice, whether a specific energy analysis should
be labeled a micro or macro analysis is not always clear.
For example, assigning labor an energy value is often viewed
as tacit acceptance of an energy theory of value. The paper
by Martha Gilliland includes a discussion of both levels of
analyses, focusing primarily on the kind of information which
energy analysis provides at each level that is not provided
by economic analysis.

In the midst of this sometimes lively controversy, the
Energy Research and Development Administration (ERDA) must
decide how best to respond to its legal mandate to utilize
energy analysis as a tool for evaluating new energy producing
technologies. Richard Williamson's paper in this symposium
volume reviews ERDA's position. He says ERDA's use of
energy analysis is almost entirely at the micro level where
energy accounting considerations are viewed as one of the
factors useful in guiding the research aimed at developing
new technologies. At present, ERDA's approach to macro work
is to support research on refinements of energy analysis to
test its use as a policy informing tool.

In summary, energy analysis is presently being used to
understand and improve how, where, and when we use energy,
specifically energy in the form of fuels. It is widely
accepted as a useful endeavor in that context. The use of
macro energy analysis for the purpose of improving economic
planning remains controversial and is not widely used. The
theory which underpins macro energy analysis is testable;
empirical data is now accumulating in the literature. While
the validity of the theory remains an open question, the
controversy is clearly beneficial in motivating both proponents and critics.

References

1. B. Hannon, Environment 14, 2,14, (1972); R. A. Herendeen, Energy Costs of Goods and Services, 1963 (Center for Advanced Computation, University of Illinois, Urbana, Illinois, 1972).

2. B. Hannon, Energy Growth and Altruism (Limits to Growth Conference, Houston, Texas, 1975); C. W. Bullard, Energy Conservation through Taxation (Center for Advanced Computation, University of Illinois, 1974); C. W. Bullard and R. A. Herendeen, Proceedings of Electrical and Electronics Engineers, Inc., 63, 484 (1975); D. A. Pilati and R. P. Richard, Total Energy Requirements for Nine Electricity-Generating Systems (Center for Advanced Computation, University of Illinois, 1975).

3. C. A. Berg, Technology Review 76,15 (1974).

4. American Physical Society, Efficient Use of Energy: A Physics Perspective (American Physical Society, New York, New York, 1975).

5. M. Slesser, Tech. Assess. 2,201 (1974); ibid., Nature, 254, 170 (1975); ibid. Science, 196, 259 (1977).

6. R. S. Berry and M. F. Fels, Bull. At. Sci. 29, 11 (1973); R. S. Berry and H. Makino, Tech. Rev. 76, 32 (1974); R. S. Berry, T.V. Long, H. Makino, Energy Policy 3, 144 (1975).

7. H. T. Odum,"Energetics of World Food Production", in The World Food Problem (Report of President's Science Advisory Committee Panel on World Food Supply, White House, Washington D.C., 1967).

8. H. T. Odum, Environment, Power and Society (Wiley-Interscience, New York, New York, 1971); H. T. Odum and E. C. Odum, Energy Basis for Man and Nature (McGraw-Hill, New York, New York, 1976).

Energy Analysis:
The Kinds of Information
It Provides Policymakers

Martha W. Gilliland

In meeting its energy policy responsibilities, the
Congress has identified a need for more adequate policymaking tools for evaluating and comparing energy alternatives.
One promising new tool currently receiving a great deal of
attention is energy analysis. The purpose of this paper is
to identify and discuss the kinds of significant policy
questions that can be answered using energy analysis, questions that cannot be answered using other kinds of analyses.

Energy Analysis

The general objective of energy analysis is to analyze
the energy flows inherent in the production of goods and services. Over the past several years, the greatest interest
has been in examining the energy costs of producing energy;
that is, the greatest interest has been in net energy analysis. As first proposed by Odum (1) and discussed by
Gilliland (2), net energy analysis measures the energy cost
of finding, producing, upgrading, and delivering a fuel to
consumers. Energy is consumed both directly and indirectly
in these processes, directly when energy inputs such as the
electricity used in uranium enrichment are required and indirectly when energy consuming material and/or service inputs
are involved. These energy inputs or subsidies are shown in
the conceptual scheme presented in Fig. 1. Although measurement procedures differ among energy analysts, they generally
agree that it is these direct and indirect energy subsidies
which are to be measured and compared to the energy delivered to the consumer.

The information produced by energy analysis can help to
inform policymaking by:

    1. measuring the impact of a policy on overall energy
        consumption, specifying the consumption changes
        that can be anticipated in energy types, quanti-
        ties, and rates;

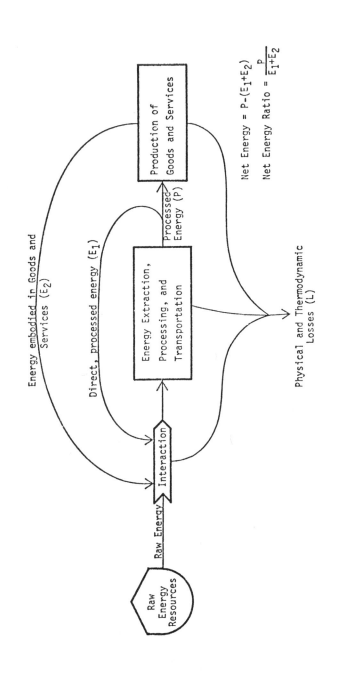

FIGURE 1: ENERGY FLOWS CONSIDERED IN A NET ENERGY ANALYSIS.

2. comparing the energy costs associated with alternative energy production technologies;
3. identifying process changes that would either increase or decrease energy consumption;
4. providing a measure of the impact of process alterations on production rates and energy consumption;
5. identifying fuel substitution opportunities and their likely impact on production rates and energy consumption;
6. sharpening policy analysis by separating energy costs which are fixed by physical laws from economic costs which are responsive to such things as changes in demand, taxes, and government regulations.

Some energy analysts, most notably Odum and Odum (3), believe that energy analysis has broader implications as a long-term planning tool. Post embargo economic experiences and theoretical considerations of energy flow have motivated research to determine the extent to which net energy controls economic feasibility and economic activity. In contrast to the conventional assumption that the rate of energy production depends on economic activity, this approach assumes that aggregate economic activity depends upon the rate of net energy production. While energy analysis will not measure or predict short-term fluctuations in demand and market forces caused by fads, fashions, preferences, and fiscal or monetary policies, it may be possible to use it to forecast longer term limits on what is economically feasible. Odum and others maintain that constraints imposed by physical laws governing energy flow can be used to establish what is economically feasible and, therefore, to forecast demand. If this relationship between energy and economics can be established, energy analysis can be a useful long-term planning tool, both for setting policies and for making economic predictions.

Energy Analysis and Economic Analysis

Some economists have reacted negatively to the introduction of energy analysis as a policy analysis tool. Their hostility is generally based on the conviction that economic analysis can provide policymakers everything that they can expect to obtain from energy analysis and more. But, the thesis underlying this paper is that energy analysis can provide answers to questions unanswerable when only economic analysis is used, and, likewise, that there are questions best answered by economic analysis. What is argued here is that policymakers will be equipped to make better informed

policy choices when the results of both kinds of analysis
are available to them. In this section, the respective
policymaking contributions of energy and economic analysis
are discussed.

As noted earlier, energy analysis measures external
energy inputs, all of which must be purchased. In a systems
sense, the energy that is purchased and the dollars used to
purchase it flow in opposite directions (1). While both
direct and indirect energy flow into the process, money used
to purchase energy inputs flows out. An economic analysis
of energy processing steps measures this money flow. Although there is a one-to-one correspondence between these
energy inputs and dollar outlays, energy and economic
analysis can diverge and lead to different conclusions con-
cerning relative costs. This is because different values
are being measured, one in energy units, the other in
dollars. For example, an energy analysis might indicate
that the energy costs of electric power generated by a coal
fired facility are less than from a nuclear generating
facility. An economic analysis might indicate that the re-
verse is the case, when costs are measured in dollars. I
discuss three factors which can cause a divergence in the
two kinds of analyses.

The Effect of Government Policies

Economic analysis necessarily includes the economic
impacts of such things as federal and state regulations, tax
incentives, and other policies on the cost of materials and
processes. That is, economic costs respond to interventions
in the market place. Energy costs, on the other hand, remain constant. For example, the severance tax on coal
leaving Montana clearly affects the economic costs of
Montana coal; it does not affect the energy costs incurred
to produce the coal. Likewise, the energy costs of uranium
enrichment are the same whether the facility is operated by
government or private industry; but economic costs may be
considerably different with different operators.

Consider the following examples which highlight the
different but complementary kinds of information yielded by
energy and economic analysis.

1. To pump one barrel of oil from a stripper well can
    require 10 thousand cubic feet of natural gas
    (converted to electricity). When oil is $11 per
    barrel and natural gas is $.50 per thousand cubic
    feet, this process returns an economic profit of
    $6.00 per barrel produced (4) but operates with an
    energy loss of 4.2 million Btu's (5). This is due
    to the disparity between the regulated price of

natural gas ($0.50 per million Btu's) and that of oil ($1.90 per million Btu's).

2. Berry, et al. (6) calculated the energy consumed in the manufacture of polyethylene film in the U.S. and in the Netherlands. While the Netherlands consumes 42 million Btu's to manufacture one short ton of polyethylene film, the U.S. consumes from 100 to 144 million Btu's for each ton manufactured; the same end product results, but U.S. energy consumption is higher by a factor of 2.4 to 3.4. Yet the economic costs of polyethylene film in the two countries do not differ greatly. There are two reasons for the difference in energy costs. First, U.S. manufacturers use natural gas as the starting resource while the Netherlands' manufacturers use crude oil, and it requires more Btu's of natural gas than crude oil to make ethylene (the intermediate product). But the price of natural gas is regulated lower than the price of crude; thus, it is still economically feasible. Second, some U.S. manufacturers use a method which is particularly energy wasteful because it calls for more process steam (7). Using this wasteful method results in the high U.S. value of 144 million Btu's.

This combination of energy analysis and economic analysis yielded two kinds of information. First, the effect of the regulated natural gas policy on energy consumption in these processes was measured. Second, two opportunities for reducing that consumption -- changing to crude oil as the starting resource and changing to the less energy consuming method -- were identified. Similar regulatory distortions exist in the generation of electricity from coal or uranium. These discontinuities and their impact on energy consumption can be identified for policymakers when energy analysis is used in conjunction with economic analysis.

The Effect of Demand

A second divergence between economic and energy analysis can occur because of the effect of demand on dollar costs. Generally speaking, dollar costs can be expected to increase when demand is high, while costs in energy units can be expected to remain the same regardless of changes in demand. People's preference for commodities cause the demand for them to fluctuate but the costs in energy units of producing the commodity are not necessarily affected by that preference. However, two demand related factors can cause energy costs to change over longer periods of time.

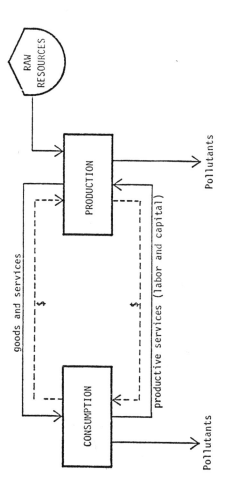

FIGURE 2: MONEY (DOTTED LINE) AND ENERGY (SOLID LINE) FLOW.

The demand for energy will affect the dollar cost of energy, which in turn may result in the introduction of energy saving process changes. In the U.S., post-embargo energy price increases have resulted in improved energy efficiencies in several industries, thereby lowering energy costs per unit of output. Secondly, if the demand for a commodity such as aluminum becomes so high that lower grade aluminum ores are tapped, the energy cost of producing aluminum will increase. Similarly, and as is clearly the case today, the demand for energy itself is so great that lower grade raw energy sources are being tapped, causing the energy cost of producing energy to increase. This increasing energy cost of tapping lower grade raw resources illustrates the utility of energy costs as an indicator of economic feasibility. In order to clarify this concept, however, a difference between economic and energy analysis theory must be examined.

The Effect of Economic Externalities

Energy flow sometimes occurs in the absence of money flow. Consider a typical economic textbook representation of money flow and GNP. In Fig. 2, the real GNP is the dollar flow of goods and services from the production box to the consumption box or the equivalent flow of wages, interest, and profit from the consumption box to the production box. The energy value and dollar value of this same flow can differ for the reasons discussed above. These differences can occur even though the economist and energy analyst are analyzing the same commodities. However, two types of energy flow shown in Fig. 2 are not included in the typical economist's representation of money flow and GNP. These are the raw fuels (such as coal in the ground or solar radiation) and the pollutants generated when the fuel is consumed (such as the waste heat). Quantities of energy are clearly associated with coal in the ground, solar radiation, and waste heat. On the other hand, the economic value of coal in the ground is measured by economic rents for land or mineral rights. Solar radiation has no economic value at the present time. In the future, "sun rights" may have to be purchased in which case it would acquire some dollar value. Up to a point, pollutants are degraded by the environment at no economic cost. But even at these pollution levels, energy is expended when the environment degrades pollutants. Until recently, the economic value of this energy expenditure was zero; it represented a free good because the supply of environmental work so exceeded demand. Recently, however, pollutant levels have exceeded the capacity of the environment to degrade them and economists have attempted to place a value on them. But they usually treat them as externalities since there are no direct exchanges of money for this work. On the other hand, energy analysts can measure this degradation of pollutants directly in terms of the energy flows involved.

The point is that the typical money and GNP diagram shows only part of the system of energy flow. Sometimes dollar values are assigned to items outside the diagram (externalities) and sometimes they are not. When dollar values are assigned, they reflect the value society places on them at some point in time, but they say little about how those values might change with time. The energy value assigned to raw resources and pollutants reflects their thermodynamic energy value; the thesis being that their thermodynamic value ultimately (over time) determines the way society will value them and the economic feasibility of producing them. As pointed out by Georgescu-Roegen (8), the flow of money as represented in Fig. 2 is essentially a closed, circular, timeless process (except for injections from the Federal Reserve Board), but the flow of the energy embodied in goods is open and one way; it is neither reversible nor timeless (9). Economic measures usually do not account for this irreversible flow of concentrated raw energy into degraded pollutants, while net energy analysis is concerned with precisely that. It is precisely this irreversible flow which may set the bounds on economic feasibility and ultimately on people's preferences.

This represents a fundamental difference between energy and economic analysis in the way raw resources and pollutants are valued. Several examples can clarify what this thermodynamic value is and what it means for policymaking. First, consider the economic production function as a general case. With regard to long-term economic models, most economists agree that labor, capital, and natural resources are the primary interacting factors which determine economic growth as measured by the real GNP. Technology, education, and other social factors are generally thought to be less important in that interaction. All of these can be considered limiting factors to economic growth. The degree to which they impose limits varies.

In practice, however, contemporary economists tend to place primary emphasis on the interaction of capital and labor, and natural resources become secondary factors like the social factors. While both primary and secondary factors have at times and under some circumstances been considered, a simple production function for the entire economy usually relates output only to the stock of capital and to the stock of labor. Indications are that these two quantities have been the dominant limiting factors since about 1900. The reason natural resources have been all but

ignored is that, given their abundance, they have not been a
limiting factor. Generally, this still appears to be the
case given the nation's enormous coal reserves and renewable
energy from the sun.

The point that is not recognized, even when natural
resources are considered, is that it is not natural resources per se that limit production; rather, it is limitations associated with their concentration, purity, and accessibility, in short their net energy. The lack of concentrated, pure fuels (not a lack of fuels) may well be the
dominant overall limiting factor to economic growth, at
least through the year 2000. It is an observable fact that
more dilute, dirtier fuels are now being used, and it appears almost certain that they will be used even more in the
future. For example, solar radiation is diffuse, large
energy expenditures are required to clean coal, and the nuclear fuel cycle is energy-intensive. Oil and natural gas
reserves located offshore and in Alaska are less accessible
than are resources located onshore in the lower 48 states.
Moreover, onshore oil and gas resources now often require
the use of energy consuming enhanced recovery technologies.

If future sources of raw materials being used are, in
fact, less concentrated and less pure than the sources used
in the past, the physical laws of thermodynamics dictate
that more energy will be expended to upgrade them. Some ob-
servers believe that technological advances will be more
than adequate to offset the affect of declining resource
quality. While vast improvements in technology are possible (improvements which can reduce energy consumption per
unit output), there is a thermodynamic limit to such advances, a limit that can be identified.--see forward. In
terms of Fig. 2, more of the gross energy produced must be
dissipated in the production box.

Consider two specific examples. The production of 50
Btu's of crude oil onshore in the 1950's required about 1
Btu in production energy costs. The production of 50 Btu's
of synthetic crude from oil shale or coal may require about
8 Btu's. The net yield for onshore oil is 49 Btu's, while
that for synthetic crude is 42 Btu's, a difference of 7
Btu's no longer available to produce goods for final demand.
Economic incentive programs for oil shale development or
coal liquefaction will not change this fact. The present
price of alternative energy sources may make synthetic crude
production economically feasible, but that price does not
change the energy costs of producing synthetic crude either.
The energy costs of making gasoline from oil shale were essentially the same in 1950 as they are now. How is it possible to make more goods for final demand in the face of this
decline in net energy? Economic predictions imply that we

can, because they recognize only that price incentives will
cause more gross amounts of energy to come out of the ground,
not recognizing that there is still a decline in the net
amount of energy available for final demand or for making
goods for final demand.

A current choice for petroleum production--offshore oil
vs. synthetic crude--provides another illustration. Assume
1000 Btu's is to be allocated to the production of crude oil
and assume that producing 25 Btu's of offshore oil requires
1 Btu in production and producing 25 Btu's of synthetic
crude requires 4 Btu's in production. The return on the
1000 Btu's investment is 25,000 Btu's of offshore oil or
6250 Btu's of synthetic crude. Obviously the former yields
more energy to produce other goods. That 1000 Btu's might
also be allocated to making devices which improve the efficiencies with which oil is consumed (e.g., housing insulation
or different car engines). Those efficiency improvements
might result in 30,000 Btu's of energy saved annually and
now available to produce other goods. While a decision
might be to do all of these (because, for example, the actual amount of offshore oil on the continental shelf is uncertain), the energy costs of the choices and the impact of
those energy costs on economic activity ought to be
analyzed.

The conclusion which emerges is that economic growth is
coupled to net energy growth, but that economic costs do not
always reflect energy costs. Furthermore, they cannot and
should not if those economic costs are to accurately reflect random preferences, fads, fashions, scarcities, and
so on. But it is impossible to rely on a measure which includes these short-term factors and still see beyond them.
Economic costs do very little by way of predicting changes
in scarcities or in analyzing limitations. All of which
tends to support the prior observation that energy flows can
be used to bound what is feasible. As such, economic predictions might be improved if the changing difference between gross and net energy production were to be included in
predictions of the production function. That is, if the
decline in the amount of concentrated fuels available for
production was accounted for; better, more accurate economic
predictions could be made.

The current dilemma is one of how to increase net
energies in the face of utilizing lower quality raw fuels.
There are some opportunities and energy analysis is the tool
which can identify them.

Opportunities for Increasing Net Energies

Increased energy requirements for production (declining
net energies) can result in fewer final goods, a situation

which could limit economic growth. We consider two possibilities for increasing or maintaining overall net energies in
the face of the lower quality domestic resources now available: (1) through technological change or advances, increase the efficiencies of the tasks which consume present
fuels and/or (2) continue to import more high quality foreign oil. Energy analysis can identify the specifics of how
to do both.

Efficiency Increases

Several researchers, most notably the Ford Foundation
Energy Program (10) and the American Physical Society (11),
have pointed out that efficiency improvements can partially
decouple growth in energy and in the economy. There are
significant potential improvements in the way machines and
equipment use energy and deliver work. There appear to be
two kinds of opportunities. The first involves simply
"tightening up" the present devices (e.g., increasing gas
mileage using the same type of car engine, insulating homes
more effectively, sealing leaky industrial steam lines).
The second involves changing the device and sometimes the
fuel type used for a given task. This latter opportunity
has only recently been discussed. It is derived from theoretical considerations of the thermodynamic concept of
energy availability (first introduced by Gibbs in 1878; see
Gibbs (12). Berg's (13) paper stimulated an investigation
into the opportunities; the American Physical Society (11)
recently researched some of the possibilities; and Commoner (14) has expanded the idea in a more popularized
series of articles.
It is possible by using textbook values to calculate
the minimum amount of energy (15) which, due to thermodynamic constraints, must be used to perform any task. A
great deal more than the minimum energy is always consumed;
energy analysis of any production or consumption step tells
us how much is consumed and where in the process it is consumed. By comparing actual consumption to the minimum required consumption, opportunities for system improvements
can be identified. However, it is neither possible nor de-
sirable to make improvements which reduce consumption to the
theoretical minimum. Such a reduction would cause processing to proceed at unacceptably slow rates. Odum and
Pinkerton (16) and Odum (17) suggest some optimal rates.
Using this comparison between actual consumption and
the theoretical minimum, the American Physical Society (11)
analyzed energy-consuming tasks in the home, the operation
of an automobile, and some industrial processes; Berry et al.
(18) analyzed energy consumption in the production of automobiles and in packaging goods; and Hannon (19) identified

opportunities for substituting labor for machines and analyzed the impact of that substitution on productivity and
income.

Several examples from space heating should clarify this
point. Natural gas furnaces are generally said to operate
with an efficiency of 75 percent (it is sometimes as low as
45 percent), meaning that 75 percent of the energy content
of the gas is delivered as heat. Electric resistance
heaters typically operate with 90 to 95 percent efficiency,
that is, 95 percent of the electrical energy is delivered as
heat. At best, electric resistance heating is 30 percent
efficient if the energy content of the coal at the power
plant is compared to the heat delivered in the home. Solar
heating can deliver, as space heat, 30 percent of solar radiation or 80 percent of the heat collected by the solar col-
lector. These efficiency expressions are known as "first
law efficiencies" for some of the devices and groups of devices which deliver space heat. The "second law efficiency"
concept of Berg and the American Physical Society compares
these devices to the best possible system (best in that it
delivers the most work) for space heating. The concern is
not necessarily with improving furnaces, but with replacing
the furnace and/or fuel type with the best possible system
for doing the job.

Second law efficiencies depend to a great extent on how
the energy source (e.g., fuel, electricity, or heat) is
matched to the requirements of the task (e.g., mechanical
work, space heat). When high quality energy (e.g., electricity) is used to perform high quality tasks (e.g., operate a
motor), second law efficiencies do not differ greatly from
first law efficiencies. But when high quality energy (e.g.,
electricity) is used for low quality jobs (e.g., space heating), second law efficiencies are much lower than first law
efficiencies. It makes no sense to heat steam to 1200°F in
a power plant only to use the resulting energy to heat a
house to 70°F.

Second law efficiencies are quite low in this country.
The national average for space heating is about six percent,
for water heating three percent, for air conditioning five
percent, and for industrial use of process steam 25 percent
(American Physical Society). Clearly, there is room for improvement. Energy analysis of these tasks can identify
wasteful steps in processes which produce energy, produce
manufactured goods, or consume energy.

One caveat on this relatively straightforward concept
is in order. Replacing "wasteful" devices with energy saving systems requires capital expenditures of energy for manufacturing. Waste during consumption can be reduced, but
only at the expense of increased energy expenditures to

construct the energy saving device. In other words, the net
energy delivered by a new energy conserving system is not
necessarily greater than the original system. Very few
researchers have calculated the energy cost of improving
second law efficiencies. Those studies that have been done,
however, indicate that more energy is usually saved than is
required to convert to efficient "machines". For example,
Putnam (20) analyzed the energy costs and savings of housing
insulation and found, for average U.S. conditions, that four
Btu's of energy were saved annually for each one Btu invested in insulating under-insulated homes.

Energy analysis can also identify limits on such efficiency improvements and, therefore, indicate limits on economic growth. Slesser (21) presents a clear example of such
an application of energy analysis. He notes that the energy
required to manufacture ammonia has decreased exponentially
since its first synthesis in 1912. But beginning about
1960, that exponential curve began to flatten as it approached the thermodynamic minimum amount of energy required
to manufacture ammonia. The thermodynamic minimum sets the
ultimate limit on the technological advances which are possible. Because processing proceeds at infinitely slow rates
when operating at the minimum, the limit to improvements or
advances in technology will always be somewhat higher than
the thermodynamic minimum.

Gyftopoulos et al. (22) estimated the thermodynamic
minimum energy requirements of a number of industrial processes in the U.S. and found that iron making uses 4.16
times the thermodynamic minimum while paper making uses 170
times its minimum. By concentrating efficiency improvement
efforts on wasteful systems, enormous energy savings are
possible. But, there is a limit and energy analysis can
identify the limit.

The Import Question

National policy goals with regard to oil imports are
complicated by foreign policy considerations, but it is still
important to know what energy analysis has to say about imports. The flow of money for imported oil is out of the
country, while money associated with the conversion of coal
and oil shale to oil remains in the U.S. economy. Both can
affect GNP, but for different reasons. One is strictly a
net energy problem in that more energy is consumed in production and, therefore, less is available for the production
of final goods, and the other is a balance of payments
problem.

While it is universally understood that oil imports are
energy imports, it is sometimes forgotten that exports of

goods, services, and technological information are exports
of energy. For example, a nuclear reactor or enriched
uranium have very high energy contents because of the years
of energy-subsidized technology that has gone into them.
Wheat exports "contain" energy in the form of tractor fuel,
fertilizer, pesticides, and highly developed technology.
This raises a question of how best to use domestic energy
sources. Domestic energy can be used to upgrade dirty, inaccessible energy sources (produce oil from coal or oil
shale), an option which would decrease net energies and probably decrease the amount of final goods produced, but would
provide an opportunity to reduce oil imports. Or domestic
energy can be used to grow more wheat by bringing more land
into production. The wheat can be exported in order to
balance the cost of continued oil imports. There may not be
enough domestic net energy to do both.

Bringing more land into food production will require
large amounts of energy subsidies, but if those subsidies
are less than those for making oil from coal or oil shale,
then, from a net energy point of view, it is better to do
the former, export the food, and continue to import oil (3).
This option increases net energies and deals with the balance of payment problem, but it causes oil imports to rise.
The risk of another embargo remains a major problem. Numer-
our proposals for protecting against an embargo have been
made, for example, oil storage. Researching and developing
dilute energy sources as substitutes for imported oil can
also be viewed as one of several, costly hedges against
another embargo. But as long as the imported oil is available, even at $11 per barrel, the economy might be better off
to use it with our own high quality fuels to make commodities for exports, then to use our own fuels to upgrade
dilute energy sources.

Summary

All raw energy sources require some external energy
inputs to make them available. Just how much varies as a
function of the accessibility and concentration of the raw
resource. Net energy measures the quantity of external
energy required to deliver the energy source to the consumer;
it includes direct energy inputs as well as the energy embodied in materials and services. Economic analysis of
alternative energy supply options can differ from energy
analysis of those same options for at least three reasons.
Economic costs include the effects of regulatory and incentive policies, respond to short-term demand changes caused
by changes in people's preferences, and treat some energy
flows as externalities. Energy costs do not include or

respond to regulatory policies or short-term demand changes, and energy costs internalize all energy flows. Energy costs reflect only the physical quantities of energy involved regardless of the particular extant market conditions and regardless of the social utility of a commodity at a point in time. As a result of these differences, two broad categories of information are provided by energy analysis; which are not generally provided by economic analysis: information about energy consumption patterns and about limits on economic activity.

First, since energy analysis tracks energy flows; it identifies and predicts energy consumption patterns which include quantities of energy, kinds of energy, and rates of consumption. In short, energy analysis examines the energy impacts of existing policies or policies under consideration. For example, energy analysis can measure the energy cost of present electric power rate structures and of proposed changes in them; energy analysis can calculate how much energy will be required to bring new technologies on line and when (and if) the new technologies will pay for their energy investment; energy analysis can measure the energy cost of insulating homes or manufacturing different kinds of automobile engines, measure the energy saved by these, and the timing of these costs and savings; and energy analysis can compare the energy impacts of the present mix of energy producing technologies to alternative mixes. In addition, energy analysis and economic analysis used together can identify policy problems (the points where the analyses diverge) and thereby identify regulations and incentives that may be exerting unintended influences.

Second, energy analysis provides information about economic activity. Energy analysis theory says that there are some physical (thermodynamic) constraints on economic activity. These constraints are not usually recognized by economic analysis because the changing quality of the raw fuels, which are required for production, is not accounted for in dollar units or in analyses of the production function. Energy analysis accounts for these constraints by evaluating changing net energies. As such, energy analysis or the incorporation of changing net energies into economic analysis can provide information on the overall level of economic activity to be expected, given a particular energy mix. Likewise, the effects of alternative energy mixes on economic activity can be evaluated. While energy analysis will not predict or evaluate fluctuations in individual industrial sectors within the economy, because it can not respond to consumer preferences; it can evaluate the overall level of activity to be expected. The implication is that overall economic activity has an energy basis.

Economic stimuli (for example, printing more money), which do not have energy to support them, are artificial (3) and may alter the internal distribution of economic activity, but not its overall level.

Efficiency changes, particularly where fuel quality is matched to the quality requirements of a task, offer an opportunity to maintain and possibly increase overall net energies, in the face of the utilization of less concentrated raw energy sources. A second possibility for maintaining net energies may be to continue importing oil and use it to make high quality exports. This latter possibility is less certain because so little is known about the energy embodied in our exports. In any case, our thesis is that, in a nation whose high quality energy is scarce, net energies must increase if the production of final goods is to increase, that energy analysis can identify and analyze the opportunities for increasing net energies, that economic analysis by itself cannot identify or analyze those opportunities, and that simply upgrading more and more dilute, inaccessible, dirtier raw resources will not increase net energies.

Regardless of the multiplicity of factors considered, be they economic, energy, environmental, political, and/or social or the emphasis placed on each, decisions would be better informed if the kinds of information provided by energy analysis were considered.

References

1. H.T. Odum, Environment, Power and Society (Wiley-Interscience, New York, New York, 1971); H.T. Odum, Ambio 2, 220 (1973).
2. M. W. Gilliland, Science 189, 1051 (1975).
3. H.T. Odum and E.C. Odum, Energy Basis for Man and Nature (McGraw-Hill, New York, New York, 1976).
4. (1 barrel x $11 per barrel) - (10 thousand cubic feet x $0.50 per thousand cubic feet) = $6.00.
5. 5.8 million Btu's oil produced - 10 million Btu's of natural gas required to produce it = -4.2 million Btu's.
6. R.S. Berry, T.V. Long, H. Makino, Energy Policy 3, 144 (1975).
7. The metal oxide catalyst method calls for more process steam than the Ziegler transition metal halide catalyst method (6).
8. N. Georgescu-Roegen, The Entropy Law and the Economic Process (Harvard University, Cambridge, Mass.,1971).
9. On earth, "non-renewable" resources (e.g. coal and oil) are constantly being created by the sun's energy but the speed with which they are made is orders of

magnitude slower than the speed with which we are
using them.

10. Ford Foundation Energy Policy Project, A Time to Choose
(Ballinger Publishing Co., Cambridge, Mass., 1974).

11. American Physical Society, Efficient Use of Energy: A
Physics Perspective (American Physical Society, New
York, New York, 1975).

12. J. W. Gibbs, The Collected Work of J. Willard Gibbs
(Yale University Press, New Haven, Conn., 1948).

13. C. A. Berg, Technology Review 76, 15 (1974).

14. B. Commoner, New Yorker 51, 38 (1976).

15. Actually free energy in the thermodynamic sense.

16. H.T. Odum and R.C. Pinkerton, American Scientist 43,
331 (1955).

17. H.T. Odum, in Ecosystem Analysis and Prediction, S.A.
Levin, Ed. (Society for Industrial and Applied
Mathematics, Philadelphia, Pa., 1975), p. 239.

18. R.S. Berry and M.F. Fels, Bulletin of the Atomic
Scientists, 29, 11 (1973); R.S. Berry and H. Makino,
Technology Review, 76, 32 (1974).

19. B.M. Hannon, Energy Growth and Altruism (Limits to
Growth Conference, Houston, Texas, 1975).

20. E.E. Putnam, Energy Benefits and Costs: Housing
Insulation and the Use of Smaller Cars (Center for
Advanced Computation, Urbana, Illinois, 1975).

21. M. Slesser, Nature, 254, 170 (1975).

22. E. P. Gyftopoulos, L.J. Lazaridis, T.F. Widmer,
Potential Fuel Effectiveness in Industry (Ballinger
Publishing Co., Cambridge, Mass., 1974).

23. I especially thank I. White for his editorial work on
the manuscript and for many insights into policy
applications of this tool. D. Kash, W. Mitsch, and
E. Rappaport made many helpful suggestions.

About the Author

Clark W. Bullard III is an associate professor at the Center for Advanced Computation at the University of Illinois, Urbana. He has published many reports and journal articles on energy demand analysis, modeling and error analysis.

2

Energy and Employment Impacts of Policy Alternatives

Clark W. Bullard, III

This paper summarizes results of energy and employment impact analyses performed by the University of Illinois Energy Research Group during the period 1970-1977. Results of net energy analyses are also included, and several critical methodological issues are discussed.

Introduction

In this paper, I would like to discuss energy analysis, an area broader than net energy analysis. I shall present a variety of results obtained during the last seven years by the University of Illinois Energy Research Group, and discuss how these results have found their way through various channels into the policy process.

Energy analysis is important for a number of reasons, both theoretical and practical. The theoretical reasons are based on the underlying premise that as purchasers in an economic system, we demand that the thermodynamic state of materials be changed to one which we value. We value a table, for example, because it's an arrangement of physical materials that suits our demands for today. There is a certain minimum energy requirement needed to change the state of materials from their natural occurrence in the crust of the earth into this useful form, and that minimum amount of energy is defined by the Second Law of Thermodynamics.*

To achieve these Second Law energy efficiencies, one must sacrifice other resources (e.g. capital and labor), eventually approaching an energy theory of value, which some people espouse and many reject. But the economic system is

---

* See Marc Ross (1977), elsewhere in this volume.

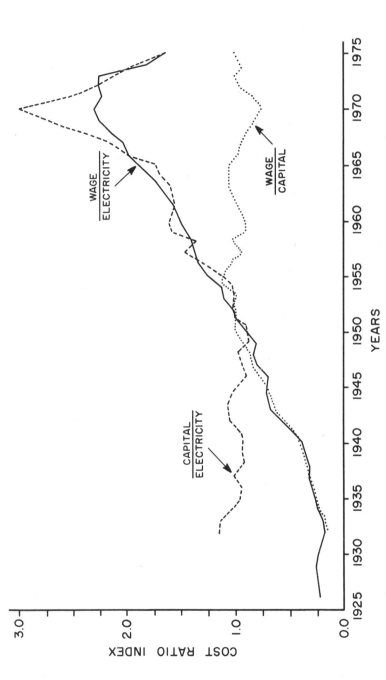

Figure 1. U.S. Industrial Labor, Electricity and Capital Cost Ratios, 1926 to 1973, Current Dollars, (Manufacturing Worker's Hourly Wage, The Industrial Price of a Kilowatt-hour of Electricity and the Yield on Aaa Corporate Bonds). Source: Hannon (1975).

operating today at a point far away from the second law minimum energy requirement for a very good reason; the values
driving the system reflect much more than simply a concern
for energy. This explains the wide variation in the energy
required produced one dollar's worth of the different goods
and services making up the GNP.* It is the nonsubstitutabil-
ity of energy that draws our attention to it from a theoretical basis, that is why Professor Odum's diagrams show energy
flows and not the carbon cycle.**

Today conventional energy resources are becoming quite
scarce and unemployment is high, so it is also important for
practical reasons to consider the energy and employment impacts of policy decisions. Warranting special concern are
those policies directed at alleviating energy or unemployment
problems. It is from this direction that we at Illinois have
approached energy analysis. In this report I shall summarize
the types of practical questions that most of our analyses
during the last six or seven years have addressed.

Figure 1 shows the relative prices of three factors of
production, energy, capital, and labor, during the last fifty
years. Note the post-embargo behavior; relative prices have
returned to levels that prevailed during the early 1960's.
Electricity prices were used because in the industrial sector,
electricity is the energy form that most directly competes
with labor. These data are presented as an introduction to
demonstrate that these ratios have remained constant over
some periods of time, have maintained well defined trends,
and have undergone substantial changes that resulted in economic pressures to substitute among capital, labor, and energy resources.

Table 1 shows the variation between the amount of energy
required to produce a dollar's worth of various products.
The substantial variation reflects the relative values society places on energy relative to other factors of production
such as capital and labor.

Figure 2 is obtained through a methodology exactly par-
rell to energy analysis, that identifies the labor required
to produce a unit of various goods and services. The method
quantifies both the direct and indirect labor inputs to produce a particular goods or service. The graph depicts a
roughly hyperbolic distribution at the economy-wide level,
showing that some products have a high energy intensity but

---

*   See Herendeen and Bullard (1975) for a tabulation of the
    energy cost of various goods and services.

** See H. T. Odum (1977) elsewhere in this volume.

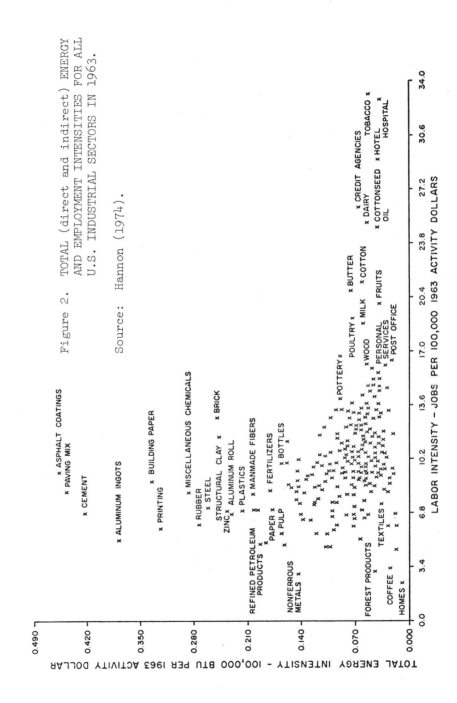

Figure 2. TOTAL (direct and indirect) ENERGY AND EMPLOYMENT INTENSITIES FOR ALL U.S. INDUSTRIAL SECTORS IN 1963.

Source: Hannon (1974).

TABLE 1. 1963 EXAMPLE ENERGY INTENSITIES (10^3 BTU/$)

PRIMARY ALUMINUM	429
FERTILIZERS	212
AIRLINES	179
GLASS	122
MOTOR VEHICLES	82
CHEESE	76
APPAREL	55
HOSPITALS	43
COMPUTING MACHINERY	32
BANKING	21
U.S. AVERAGE	82

Source: Bullard and Herendeen (1975).

TABLE 2. ENERGY PAYBACK PERIODS (see page 33)

NUCLEAR PLANT*	0.20 YR.
HOME INSULATION	0.25 YR.
OIL WELL → REFINERY → POWER PLANT	0.62 YR.
STORM WINDOWS	0.82 YR.

* Partial Fuel Cycle.

Source: Pilati (1977).

a very low labor intensity, and vice-versa. So by spending dollars on things at one end of the hyperbola instead of the other, consumers have a substantial impact on total U.S. energy demand and total employment.

Next I will describe several categories of results we have obtained, and how they were eventually used in the policy process.

Beverage Container Systems

One of the first analyses performed by the University of Illinois Energy Research Group calculated the total energy and labor requirements of two systems for delivering beverages: the throw-away container system and refillable containers.* The results indicated that a nationwide switch to returnables could produce a net increase in the number of jobs and a substantial net decrease in energy use. It takes about one-half to one-third as much energy to refill a beverage container as to remanufacture one. It is acknowledged, of course, that these two systems are not perfect substitutes. But, they are close enough to being substitutes that the systems have become the target for legislative action because of litter and other reasons. And the need has arisn for information on the economic, energy, and employment impacts. The analysis clearly identified the distribution of employment and energy impacts, and the stakeholders: unions, local franchise operators, container manufacturers, and the public at large. Utilization of these results in the policy process occurred on two levels. The first users were interest groups on both sides of the controversy, who used the results as an organizing and attention-getting tool. After several years of public debate the matter of regulating nonreturnable containers came before legislative and judicial bodies and the researchers re-entered as witnesses to articulate and clarify the issues for the decisionmakers.

Highway Construction

Another study analyzed the energy and labor impacts of diverting the highway trust fund to alternative government programs.** The results indicated that transferring funds from the highway trust fund into almost any other government activity would produce a net increase in jobs and a net decrease in energy consumption. These energy results were quite conservative, considering only the construction activity and neglecting the promotion of autos and trucks over trains through highway construction. The issue had surfaced during the energy crisis so the energy and employment were among the

---

*   See Hannon (1972) and Folk (1972).

**   Hannon (1974).

most important factors in the Congressional debate. Again, the involvement of the researchers was in two steps: first to identify the interest groups and provide the general public with basic information, then to appear as witnesses to repeat, and clarify the matter for legislative action.

Dam and Waterway Construction Projects

Another activity analyzed was the U.S. Army Corps of Engineer Construction Program: the energy and employment impacts of transferring funds from the Corps' projects to a variety of other programs. The results showed the relatively small amount jobs created by the Corps of Engineers' construction programs vs. other Federal spending options.* The results were utilized through the same two-step process outlined earlier; the public was provided with information, interest groups cited certain parts of it and generated political pressures, and then the research results were used to facilitate and justify the final decision.

Project Independence

We have also analyzed the energy, labor and capital impacts of Project Independence. Bechtel Corporation had performed detailed study of the amount of scientific and technical and other manpower needed directly to construct energy-related facilities for associated with the President's ten-year Project Independence Program.** Then our analysis identified the indirect requirements for scientific, technical and other manpower, and showed they were about three times greater than the direct requirements.*** It is this total requirement that must be compared to the available labor supply. Questions had also been raised about the rapid growth rates possibly reducing the net energy rate of return from the construction program. Our results showed that the energy returned from the construction program exceeded direct and indirect energy inputs by a factor of six: for every BTU used constructing and operating facilities--six BTU's were produced.

Energy Supply and Conservation: Net Energy Yield

Table 2 shows a few net energy results for specific energy supply and conservation options. The parameter shown is the payback period. It is important because if the program's doubling time were less than the pay-back period, the program would be a net energy sink, because all power plants

---

* Hannon & Bezdek (1974).

** Carasso (1974).

***See Bullard and Pilati (1976).

would be fully dedicated to replicating themselves.

Take for example the nuclear electric system vs. the oil wells-to-refineries-to-power plant system. Comparing these on a kilowatt-hour output basis, we obtain energy payback periods on the order of .2 to .6 years. These results account for the energy quality factor by assuming that if the energy had not been used to build the nuclear power plant, but had instead been directed through the existing fossil-fuel electric system, a certain amount of electricity would have been produced. That electricity is compared with the annual output of the power plant to obtain the payback period. Home insulation and storm windows yield the same type of result, but the energy quality comparison is straightforward because only fossil fuels are involved. Concern has been raised that a massive home insulation program might require more energy to make the insulation than would be saved in the first year or two of operation. That concern is ill-founded unless the program's doubling time approached the levels shown in Table 2.

Energy Tax Impacts

Figure 3 shows the direct and indirect energy impacts of personal consumption expenditures. The total energy required for a family's lifestyle is roughly proportional to total income, but poor people consume a much greater fraction of their energy directly (almost two-thirds), while the affluent consume most of their energy indirectly through things such as airplane tickets, "plastic" carpeting for large homes, etc. It is very important to look not only at energy use as a function of income, but also many other variables. In this manner, one could assess the impacts on various socio-economic groups of energy pricing policies, (e.g. decontrol), energy taxing policies, and the like. Roughly, the results indicate that an energy tax would be about as regressive as a sales tax, since total energy use is approximately proportional to income. This points out the need for progressive redistribution of tax revenues generated.

The Energy Balance of Trade

Figure 4 shows the energy balance of trade for the United States. The question had been raised as to whether there was enough energy to produce agricultural products that would have to be exported to pay for imported oil. The results show our exports to be less energy intensive than our imports. The energy imported both directly (e.g. oil) and indirectly (e.g. embodied in Volkswagens) has risen steadily over time. These results not only answer the immediate practical questions

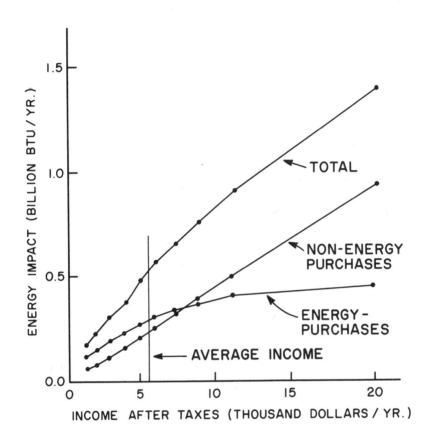

FIGURE 3. Direct, Indirect and Total Family Energy Impact
Plotted Against Family Income, 1960 to 1963.

Source: Herendeen (1974).

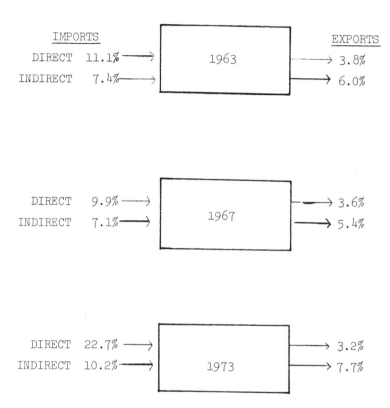

IMPORTS   EXPORTS
DIRECT   11.1% ⟶   1963   ⟶ 3.8%
INDIRECT   7.4% ⟶   ⟶ 6.0%

DIRECT   9.9% ⟶   ⟶ 3.6%
INDIRECT   7.1% ⟶   1967   ⟶ 5.4%

DIRECT   22.7% ⟶   ⟶ 3.2%
INDIRECT   10.2% ⟶   1973   ⟶ 7.7%

PERCENTAGES OF DOMESTIC PRODUCTION

Figure 4.   THE U.S. ENERGY BALANCE OF TRADE

Source: Bullard and Herendeen (1976)

regarding our energy balance of trade, but also shed light on problems in theoretical economics such as the Leontief paradox. Leontief (1953, 1956) showed that the theory that U.S. exports should be relatively more capital intensive and less labor intensive did not hold. Using energy analysis we have explained this paradox in terms of the complementarity of capital and energy and the relative shortage of energy in the U.S.*

Energy and Employment

Our Energy Research Group results on energy and employment impacts are summarized in Table 3. Results are in four categories, since both energy and labor impacts can be either positive or negative. All results are based on a constant income assumption: if the costs of two activities differ, the balance was assumed to be spent on average personal consumption expenditures.

Methodological Issues

The methods for calculating direct and indirect energy and employment requirements for various goods and services have been described in reports cited earlier. One aspect critical to net energy analysis requires some elaboration here; it is the system boundary definition.

The system boundary definition employed for all the above analyses was identical to that defining the Gross National Product. Energy resources enter the economic system when they are extracted from the earth, and "exit" the system embodied in the goods and services making up the GNP. These include personal expenditures, government consumption and gross private capital investment. This standard economic system boundary is appropriate for the class of analyses cited earlier whose purpose was to determine the energy and labor impacts of alternative economic activities or policies. Elements of the GNP are in fact the decision variables for a wide variety of problems.

The GNP system boundary has problems, too, which have become most apparent in net energy analyses. For example, calculating the total energy required to produce electricity using the GNP system boundary would count only the energy embodied in the power plant's (say) coal input plus that embodied in incidental operating inputs such as paper for billing purposes. The energy embodied in the structure of the

---

* See Bullard and Hillman (1975).

Table 3.   ESTIMATED ENERGY-LABOR IMPACTS OF A VARIETY
OF CONSTANT INCOME CHANGES IN THE STRUCTURE
OF THE U.S. ECONOMY

Source: Hannon (1976).

PROJECT	NEW JOBS PER QUADRILLION NEW BTU (SAVED)
Changing from....	
....Plane to Train (Intercity)	930,000
....Throwaway to Refillable Beverage Containers	750,000
....Car to Train (Intercity)	700,000
....Owner-Operator Truck to Class 1 Freight Train	675,000
....New Highway Construction to Health Insurance (Federal)	640,000
....Car to Bus (Intercity)	330,000
....Car to Bus (Urban)	210,000
....New Highway Construction to Personal Consumption	200,000
....Car to Bicycle	200,000
....Plane to Car	160,000
....Plane to Bus	140,000
....Electric to Gas Stove	160,000
....Electric to Gas Water Heater	120,000
....Electric Commuter to Car	110,000
....Electric to Gas Clothes Dryer	100,000
....Frost Free to Conventional Refrigerator	60,000
....Plush (25 appliances) to Moderately Equipped (16 appliances) Kitchen	30,000
....New Highway Construction to Railroad and Mass Transit Construction	30,000
....Present to Increased Home (Oil Heat) Insulation	15,000
....Moderate to Spartan (4 appliance) Kitchen	10,000

Table 3 (continued)

PROJECT	JOBS GAINED PER QUADRILLION BTU LOST (USED)
Changing from Electric Commuter to Bus	530,000

- -

PROJECT	JOBS LOST PER QUADRILLION BTU LOST (SAVED)
Changing from....	
....Black & White TV to Radio	35,000
....Present to New Electricity Supplies	75,000
....Bus to Bicycle	330,000
....Car to Motorbicycle	430,000
....Color TV to Black-White TV	1,750,000

- -

PROJECT	JOBS LOST PER QUADRILLION BTU LOST (USED)
Changing from....	
....Beef Protein to Textured Soy Protein	720,000
....Beef Protein to Direct Bean Consumption	860,000
....Beef Protein to Complete Soybean Meat Analog	970,000
....Class 1 Truck to Container Train	13,600,000

power plant itself would be ignored!* This anomoly occurs
because the plant is an independent capital investment -- part
of GNP -- and not as an input to electricity production. To
correct this problem, Bullard, Penner and Pilati (1976) have
calculated the energy cost of goods and services with capital
investment inside the system boundary. This effectively increased the energy intensity of all goods and services about
16% on the average.

Similar arguments have been made regarding certain government research, development, and regulatory activities. For
example should the energy required directly and indirectly
for the activities of the Nuclear Regulatory Commission be
counted as an input to nuclear electricity production rather
than an independent output of the economic system? (It is
actually counted as GNP.) If the nuclear industry were self-
regulating, these costs would in fact be included in the cost
of nuclear power, and the energy inputs would also be counted.
These effects are difficult to quantify; particular problems
arise in the allocation of R&D activities over various products and overtime. The effects for nuclear and fossil-elec-
tric production have been estimated by Amado (1977) and found
to be substantial, but not enough to render either technology
a "net energy sink."

By far the most important issue related to the system
boundary definition lies in the treatment of labor inputs.
Some claim that the energy consumed by employees travelling
to work should be assigned to the product they make.** For
example, energy consumed to build new towns near new energy
resource developments would be considered as an input to a
net energy analysis of that resource. While seemingly logical with respect to the particular problem at hand, this approach is inconsistent with the conventional economic system
boundary. It is incorrect to use energy intensities derived
using the conventional economic system boundary to quantify
the energy embodied in these labor-related inputs to production process.*** The assumption explicitly employed to obtain these energy intensities was that labor inputs to production processes have zero energy cost, and the energy embodied in personal consumption goods and services are a
benefit to consumers, not a cost of production.

---

* A correction has been made for this effect in the net energy analyses cited in this paper.

** See Odum (1977) for arguments along these lines.

*** This applies to the detailed energy intensities of
Herendeen and Bullard (1975) as well as the aggregate
Energy/GNP ratio used by Odum (1977).

Conclusion

The approach to energy analysis outlined here is one where economists and engineers are in general agreement. The concerns of ecologists such as Odum are not adequately taken into account by this methodology, since it is not based on ecological system boundaries.

The economic system boundary is fundamentally anthropocentric, and recognizes consumer decisions as the driving force and independent variable in the analysis. For practical problems related to the assessment of energy impacts of economic activities, this is indeed the appropriate system boundary to use.

For certain theoretical problems other systems boundaries may be more appropriate. I believe Odum (1977) has made this point well, and is correct in recognizing from an overall ecological system viewpoint, that the energy cost of Alaskan oil includes the jet fuel required for employees to return to the lower 48 to visit their families occasionally. But I believe Odum is incorrect in quantifying such energy impacts using the results of analyses based on the conventional economic system boundary. The "correct" answer will await further research and analysis.

References

Amado, D. L., "Effects of Redefining the System Boundary on the Net Energy of Electric Generating Systems", CAC Tech. Memo. 85, Center for Advanced Computation, University of Illinois, Urbana, IL, February, 1977.

Bezdek, Roger and Bruce M. Hannon, "Energy and Manpower Effects of Alternate Uses of the Highway Trust Fund," Science, August 1974, Vol. 185, pp. 669-675.

Bullard, Clark W. and Robert Herendeen, "Energy Impact of Consumption Decisions," Proceedings of the IEEE, March 1975.

Bullard, Clark W. and Robert Herendeen, "Energy Cost of Consumer Goods 1963/67, CAC 140, Center for Advanced Computation, University of Illinois, Urbana, IL, November 1974.

Bullard, Clark W. and David Pilati, "Direct and Indirect Requirements for a Project Independence Scenario," Energy, Vol. 1, pp. 123-131, 1976.

Bullard, Clark W. and Arye L. Hillman, "The Energy Intensity of U.S. Tradeable Goods Production in the 1960's: Implications for Leontief Paradox and Project Independence," CAC 175, Center for Advanced Computation, University of Illinois, Urbana, IL, August 1975.

Bullard, Clark W., Peter Penner and David Pilati, "Energy Analysis Handbook," CAC 214, Center for Advanced Computation, University of Illinois, Urbana, IL, October 1976.

Carasso, M., et al., The Energy Supply Planning Model, Bechtel Corporation Energy Systems Group draft final report, Vols. 1-3, July 1975.

Folk, Hugh, "Two Papers on the Employment Effects of Mandatory Deposits on Beverage Containers," CAC 73, Center for Advanced Computation, University of Illinois, Urbana, IL, 1972.

Hannon, Bruce, "Bottles, Cans, Energy," Environment, Vol. 14, No. 2, March 1972, pp. 14-19.

Hannon, Bruce, "Options for Energy Conservation," Technology Review, Feb. 1974.

Hannon, Bruce, "Energy, Growth & Altruism," October 1975. 1st Prize - 1975 Mitchell Award.

Herendeen, Robert A. and Clark W. Bullard, "U.S. Energy Balance of Trade, 1963-1973, Energy Systems & Policy, Vol. 1, No. 1, pp. 383-390.

Herendeen, Robert A., "Affluence and Energy Demand," ASME Technical Digest, 73-WA/Ener-8, 1974.

Leontief, W., "Domestic Proportions and Foreign Trade: The American Capital Position Re-examined," Proceedings of the American Philosophical Society, 97, (September 1953) pp. 332-349.

Leontief, W., "Factor Proportions and the Structure of American Trade: Further Theoretical and Empirical Analysis," Review of Economics & Statistics, 38, (November 1956) pp. 386-407.

Odum, H. T., (1977) Elsewhere in this volume.

Pilati, David, "Energy Analysis of Electricity Supply & Energy Conservation Options," Energy, Vol. 2, No. 1, pp. 1-7, 1977.

Ross, Marc, (1977) Elsewhere in this volume.

Much of this work was supported by the Energy Research and Development Administration.

About the Author

Marc Ross is a specialist in energy policy and professor of physics at the University of Michigan, Ann Arbor. He publishes frequently on matters related to energy policy, most recently "Energy Efficiency: Our Most Underrated Energy Resource" (Bulletin of the Atomic Scientists, *November 1976) and "The Potential for Fuel Conservation"* (Technology Review, *February 1977).*

3

Second Law Efficiencies and Public Policy

Marc Ross

Introduction

As a physicist I approach the problem of analysis of energy systems with the attitude: what is the most important question which has to be illuminated and how can that limited question be adequately addressed? This is in contrast with the systems approach where completeness, and definition of system boundary are critical, but where often, I feel, key components of the system do not receive adequate attention.

I believe that the most important energy policy question in the next years and decades will involve the balance between investment in new energy supply and in more effective energy using systems. How much research, development and demonstration and how much capital formation should the nation undertake, for example, in better buildings (from a thermal standpoint) and in improved industrial processes, as contrasted with what it undertakes, for example, in outer continental shelf oil, nuclear fission power and synthetic fuels from coal? We can do some of all these things. But, as there are many difficulties and the costs are high, we cannot afford to do all of them thoroughly.

This broad policy question has not received much attention and is especially hard to deal with because the social and economic character of the two types of technology (more effective energy use and new energy supply) are so different. In this situation one primary need of policy makers

is tools to evaluate energy use technologies
and the potential for their improvement.

In this paper I will discuss performance
indices for energy using systems. I consider two
kinds of indices: general performance indices,
like miles per gallon, and an efficiency index.
Most of my attention is given to the latter.
Both of these indices can be stated in terms
of "net energy" but for the purposes of this
paper the discussion is simplified by assuming
that indirect inputs (e.g., fuel use implied by
material inputs) to energy using systems such as
buildings, automobiles and industrial processes
are usually small enough to be treated as correc-
tions. I think the other presentations in this
session confirm this assumption. In other words
I concentrate in this paper on the analysis of
the most obvious aspects of energy consumption,
the direct use of fuel by an energy use system.
To summarize, my primary concern is not so much
net energy analysis as a whole (the topic of the
other presentations) but a deeper understanding
of the direct energy consumption component in
that analysis.

General Energy Performance Indices

Energy performance or fuel consumption per unit
of product, is primarily determined by technology,
and will be modified mainly by technical change.
In contrast, the rate of consumption of a product
depends on other considerations, like income
and lifestyle. It would be very useful if the
energy performance of all major categories of
energy consumption could be monitored so that
technical improvements in performance could be
encouraged. The existence of convenient indices
would allow easier administration of such areas
of public responsibility as the regulation of
energy performance, financial subsidies, tax
penalties or incentives, and support of research
and development related to energy efficiency.
Such indices would also assist the consumer.
Examples of existing systematically determined
energy performance indices include miles per
gallon for different types of automobiles, as
measured by the EPA, and the EER measure, where
EER is the ratio of heat energy removed to
electricity consumed for electric airconditioners,

under conditions specified by the manufacturers'
association for that industry.

Such general energy performance measures
are not at this time systematically determined
in many important areas of energy use. Systematic
evaluation of energy performance in all important
areas would require significant effort to define
and develop the means of measurement; and significant costs would be involved in continuing
measurements.

Consider, as an example, the measurement of
miles per gallon in automobiles. This index of
energy performance is obtained by the EPA on the
basis of laboratory tests of vehicles under fully
standardized conditions. Two standard driving
regimes are now used: the Federal (urban) driving
cycle and constant high speed driving. Standardi-
zation in testing is achieved by simulating the
driving regimes with stationary equipment. It is
likely, as experience with the standardized tests
accumulates, that they will be modified to
reflect certain conditions, like tire resistance
and air resistance, more accurately. Nonetheless,
these standard tests are a great improvement over
the previous hodgepodge of tests performed by
manufacturers, magazines and individuals, which
produced widely-varying results for a given
make and model automobile.

Some problems remain with this test, however,
above and beyond simple questions of accuracy.
One is the variation of gas mileage from driver
to driver. Another is the variation in gas
mileage as a given automobile gets out of tune or
grows older. While these probelms are of interest,
they are not so critical as to render the results
of the standardized tests useless. Other kinds of
products would offer more serious difficulties.
For example, buildings are not standard products
of manufacture. The energy performance of buildings would have to be evaluated on a more or less
individual basis as well as being sensitive to
the local climate.

Efficiency

I hope it is clear that the definition of
general energy performance indices for different

technologies is an interesting topic for research. As useful as such performance indices could be, they do not indicate how much technical improvement might eventually be accomplished. In this paper I will discuss a subsidiary performance index designed for that purpose: efficiency. The general energy performance we have been discussing can be thought of as the energy consumption of an appropriately defined ideal system, divided by the efficiency of the system. Thus

Actual Energy Performance =

$$\frac{\text{Energy Use by Ideal System}}{\text{Efficiency}}$$

For example for a car

Actual Energy Use per Mile =

$$\frac{\text{Ideal Energy Use per Mile}}{\text{Efficiency}}$$

In order to properly define efficiency, however, a number of questions must be answered.

Available Work

An energy efficiency index should be a single number between zero and one. In order to develop a definition we must have a common basis for comparing quantities of energy. As I will show, comparison of energies directly can be misleading. Energy exists in two general categories: thermal and non-thermal. Non-thermal energies include forms such as gravitational, chemical, electrical and nuclear as well as energy of bulk motion. The non-thermal forms can be fully converted one into another by ideal processes or into thermal energy. Thermal energy is of lower quality; according to the second law of thermodynamics it can only be partially converted into other forms. Carnot showed, for example, that if thermal energy in the amount Q is obtained at a constant temperature T, the maximum amount of non-thermal energy which can be created is

$$W_{max} = Q(1-T_o/T)$$

Here T_o is the ambient temperature, $T > T_o$, and

both T_o and T are absolute temperatures. The quantity W_{max} is called the <u>available</u> <u>work</u> contained in the thermal energy Q. While the maximum work which can be done using thermal energy Q is less than Q by an amount depending on the temperatures, the maximum work that can be done using non-thermal energy is, in principle, equal to the quantity of non-thermal energy. Thus, for example, the available work provided by a system which produces both thermal energy Q at temperature T and non-thermal energy E is

$$Q(1-T_o/T) + E.$$

A general definition of available work, which was developed by Gibbs, is presented in chemical engineering texts in thermodynamics (<u>1</u>) and in an American Physical Society Study (<u>2</u>).

Task Definition

One further step needed to specify efficiency is the definition of task. We are interested in assessing the effectiveness of a system or device whose purpose is to perform a certain task. There is considerable flexibility in drawing a boundary around the system. For example, one possibility for the automobile - one which I think is very useful - is to focus on the engine plus drive train, and to take the vehicle's shell, total weight, and tires as given. The task of the engine plus drive train could be to supply rotational energy to the wheels of a vehicle with these given external characteristics so as to propel it around the Federal Urban Driving Cycle. Clearly the task could be defined differently. Clearly the energies in question will depend in detail on the definition of task.

Conventional efficiencies are based directly on energy, not on available work, and tend to be device or process specific rather than task oriented. In distinction, let us define the "2nd-law efficiency" of a system as the ratio of available work produced by the system in direct performance of its task, to the available work consumed by the system. These concepts and a survey of efficiencies in the economy are presented in the American Physical Society Study (<u>2</u>) and in papers by Robert H. Williams and myself (<u>3</u>).

Three Applications

1. The available work actually consumed by the engine plus drive train of a car is given by

the fuel energy consumed. An ideal thermodynamic
system could convert all that energy to rotational
motion at the wheels. So the available work
consumed by an ideal system would be equal to
the available work required by the task. This
requirement is specified by the needs to overcome
losses in the tires, air resistance and accelerat-
ing the vehicle (braking energy). The energy
output at the wheels is the same as the output of
available work because they are non-thermal forms.
As a result automotive efficiency is the same
number whether quoted as a conventional efficiency
or a 2nd-law efficiency. Typical current auto
efficiencies are 9% for the FUDC and 13% for 55
mph driving.

2.   The available work required by a building
heating system depends again on task definition –
the specification of temperatures and the heat
loss from the building. Let us assume for example
that the task is to deliver warm air into the
desired space at 30°C, in a climate with ambient
temperature 5°C. A furnace type heating system
which delivers 3/5 of the fuel energy into the
desired space would be described as having a
conventional efficiency of 60%. Since the
quantity $1-(T_o/T)$ is 0.083 in this case, the
2nd-law efficiency is 5% for this heating
system. What this means is that a different
heating system – in particular an ideal heat
pump – could perform the same task with an
input of 1/20 the amount of fuel. A related,
possibly more interesting, application of the
second-law efficiency concept might be evaluation
of a system which performed combined tasks
such as providing electricity, hot water, and
space heat.

3.   The production of high-btu gas from coal
is a flow process analogous to a refinery process.
The task is to act on input material streams to
convert them to desired output streams. (Generally
there are also waste streams.) Thus for example
if the desired products contain 95% as much
energy as the inputs including any added fuel,
the conventional efficiency would be quoted as
95%. To compute 2nd-law efficiency, the

consumption of available work* is that of added
fuel to carry out the process (i.e. the available
work of fuel which is required in addition to
the inputs fixed by materials balance). In the
case of production of high-btu gas from coal,
I assume here that the input streams are simply
carbon and water and that the process is:

$$2C + 2H_2O + A \rightarrow CH_4 + CO_2$$

where A stands for the additional available work
which must be provided to make the process go
(with liquid water and under standard conditions
of temperature and pressure). Ideally this
process requires the consumption of 7 kilo-
calories of available work per mole of methane
produced, or A_{min} is 0.036 of the available work
content of the methane. The projected convention
tional efficiencies are typically** about 60%.
That is, if we consider a quantity of methane
produced such that its available work is unity
the conventional efficiency is

$$\frac{1}{A_{act} - A_{min} + 1} = 0.60$$

Then A_{act} = 0.70 of the available work in the
methane. The second law efficiency is then
A_{min}/A_{act} = 0.036/0.70 = 5%. One policy implication of quoting a process as "5% efficient" as
constrasted with "60% efficient" is simply the
force of the numbers.

---

*There is an additional technical point that
while available work is essentially the energy
content of a fuel it differs somewhat from the
heat of combustion which is commonly quoted.
In principle available work is the better
measure of a fuel's capability.

**There are a number of processes such as
Lurgi, BI-GAS, HYGAS, COED and others, all much
more complex than suggested by the simple reac-
tion of carbon with steam which is used for
purposes of illustration here.

Policy Implications

Second-law efficiency is an index which (1), forcefully compares performance with the ideal process, and (2) is general enough to be applied to any system, any complex of processes. This means that the index, respectively, (1) focuses attention on redesign of processes, and (2) can be applied to "cascaded processes", such as cogeneration of steam and electricity, which offer some of the most important opportunities for fuel conservation. (4)

Two problems with the index suggest themselves. It has been hard enough here for me to define the index; in many cases considerable measurement and analysis would be required to determine its value. I believe this process of analysis, though costly, would pay for itself many times over in fuel savings that would develop from insights gained from the analysis. A second problem is that efficiency is only part of the story of fuel consumption. Several other aspects could be mentioned, by far the most important being the task which we want to get done. Thus, referring back to an earlier topic, fuel consumption per unit of task accomplished is:

$$\frac{(\text{fuel consumption required by an ideal system})}{\text{2nd-law efficiency}}$$

I have only mentioned the numerator in passing. In the cases of passenger transport and of heating buildings the task might be modified economically without change in comfort or convenience so as to reduce losses. Indeed the opportunities to save fuel this way (e.g. by reducing vehicle weight, streamlining vehicles, redesigning tire and suspensions, insulating buildings and tightening them against infiltration, etc.) may perhaps be more important then improving efficiency as such.

I have in addition encountered some objections to the 2nd-law efficiency index as "impractical". Thus for example, a specific type of space conditioning device may have a conventional performance index of 60% and engineers would <u>not</u> like to see their efforts to improve this to, say, 72% described as (referring to my second application above) as an efficiency improvement from 5% to 6%. I believe that if one's purview is a particular kind of device or process then the conventional index is of course valuable. (It always must be a gross simplification to describe performance of a complex system by one number). The point is that the 2nd-law efficiency index is valuable because it challenges "technology" to also consider a long-range program of improvements in fuel conservation. The policy issue it impinges on is whether we will make long-range plans for development of fuel conservation techniques. The present long-range policy of pouring money into development of exotic new energy supply technologies is half a policy at best. This policy which neglects long-term development of conservation techniques is based, in my opinion, on a misunderstanding of the energy problem.

REFERENCES

1. J.H. Keenan, Thermodynamics, John Wiley,
1948. Charles Berg, "A Technical Basis for
Energy Conservation", Technology Review, Feb. 1974,
and Federal Power Commission staff report
FDC/OCD/2, April 1974. E.P. Gyftopoulos, J.J.
Lazaridis, and T.F. Widmer, Potential for Fuel
Effectiveness in Industry, a report to the
Energy Policy Project of the Ford Foundation,
Ballinger, 1974.

2. W. Carnahan et al., "Efficient Use of Energy:
A Physics Perspective," in Efficient Energy Use,
American Institute of Physics Conference
Proceedings, Vol. 25, Aug. 1975. The study has
been summarized by Robert Socolow, Physics Today
Vol. 28, Aug. 1975.

3. Marc H. Ross and Robert H. Williams, "Energy
Efficiency: Our Most Underrated Energy Resource",
Bulletin of the Atomic Scientists, Nov. 1976,
and "The Potential for Fuel Conservation"
Technology Review, Feb. 1977.

4. Robert H. Williams, "The Potential for
Electricity Generation as a By-Product of
Industrial Steam Production in New Jersey",
Center for Environmental Studies report, Princeton
University, June 1976.

About the Author

Howard T. Odum is graduate research professor of Environmental Engineering Sciences, University of Florida, Gainesville. He has published widely during the past 20 years in the fields of systems ecology, energy analysis, biological oceanography, biogeochemistry and ecological engineering and modeling. His recent books include Environment, Power and Society *(John Wiley, 1971) and* Energy Basis of Man and Nature *(with E.C. Odum; McGraw-Hill, 1976).*

4

Energy Analysis, Energy Quality, and Environment

Howard T. Odum

Energy analysis is the modeling of systems accompanied by an evaluation of the energy flows inherent in the system. It includes a synthesis of parts into whole patterns where energy flow is used as the common unit of measure among parts. In practice, energy analysis starts with a diagram of important flows, structures, storages, and process inter-actions. Such a diagram is accompanied by numerical evaluation and appropriate tabular documentation. This evaluated energy diagram shows simultaneously energy balances, energy transformations, kinetics, material flows, information flows, and work transformations. From this basic energy diagram, various aggregate calculations, and simulations can be carried out. These result in an evaluation of the role parts of the system play in maintaining the vitality of the whole. Energy analysis shows common characteristics among systems of different types and suggests new energy concepts.

The energy flows of one type required to support energy flows in another part of the system define the energy cost of that part, and the energy cost is often a measure of the potential value of the part to the system as a whole. The quality of energy is measured by the Calories of one type that can generate a Calorie of other types, and the ratio suggests which features of the system must have large amplifier effects to justify their accumulated energy cost.

As part of the basic science of energetics and systems, energy analysis diagrams have been used for a half century in many fields to show overall relationships and resources. In recent years, as fossil fuel supplies diminish, overall environmental energy analysis procedures have become of special interest for showing the energy basis of the economy of humanity. This is a description of some of the methods of energy analysis as used both to understand the energetics of man and the biosphere and to evaluate alternative choices in energy use. The paper is divided into four parts. The

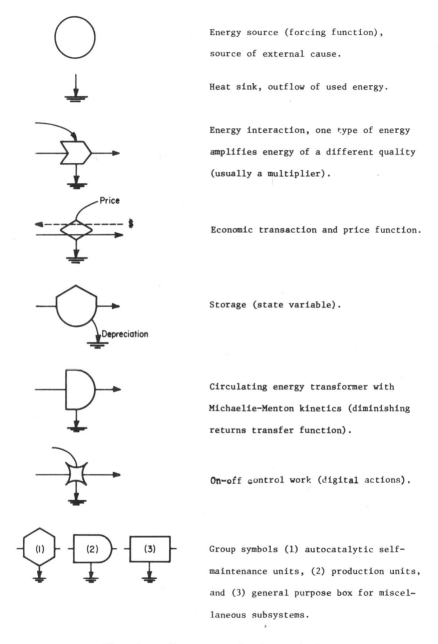

Energy source (forcing function), source of external cause.

Heat sink, outflow of used energy.

Energy interaction, one type of energy amplifies energy of a different quality (usually a multiplier).

Economic transaction and price function.

Storage (state variable).

Circulating energy transformer with Michaelie-Menton kinetics (diminishing returns transfer function).

On-off control work (digital actions).

Group symbols (1) autocatalytic self-maintenance units, (2) production units, and (3) general purpose box for miscellaneous subsystems.

Fig. 1.   Energy analysis symbols.

first section includes a description of the basic energy diagram and some of the theory which underlies data preparation.
The second section discusses the concept of energy quality,
its evaluation, and its significance as a value measure. The
third section applies the concept of energy cost and energy
quality to real world natural and economic systems. The
final section applies these same concepts to some alternatives of special interest in energy policy-making today.

Preparing Energy Analysis Data

Data in several forms are required for a full energy
analysis of the system of interest; these data are derived
from an evaluation of the heat equivalents of energy flows.
Certain theoretical factors which explain the observed patterns of energy flow in many systems aid in data preparation
and in diagramming. In this section, the energy symbols
used in diagramming are given first along with an example.
Second, the evaluated energy flows (as heat equivalents) inherent in the example are given--a first law evaluation.
Third, the maximum power theory, which may explain observed
patterns of energy flow, is introduced. Fourth, some characteristic webs of energy flow which develop because of the
maximum power theory are given. Finally, the concept of energy of equivalent quality is discussed via an energy cost
diagram.

Energy Symbols and Diagrams

Although different symbols have been used by different
authors diagramming systems for various purposes, the full
potential of energy analysis requires that the symbols carry
mathematical and energetic meaning simultaneously. For this,
the energy analysis symbols in Fig. 1 are available as used
and described in many books and papers since 1967 (1). An
energy analysis diagram of Silver Springs, Florida, is given
in Fig. 2 which shows the flows of energy of many types and
in several forms. It indicates how these flows interact as
they do work and shows all flows ultimately leaving the system as degraded heat. While Silver Springs is predominately
a "natural" system, note that its economic component is included in Fig. 2. As the diagram indicates the work of the
natural processes interfaces and attracts the flow of money
in tourist-supported developments.

First Law Evaluation; First Law Diagram

The next step after diagramming the system is a numerical evaluation of the energy flows.

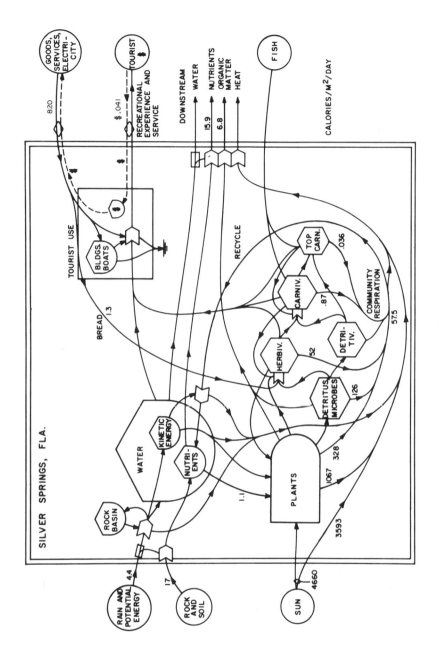

Fig. 2.   Silver Springs, Florida, example of an energy
analysis diagram evaluated with numerical values of heat
equivalents to form a first law diagram (2).

In Fig. 2 the average heat equivalents stored or flowing per time are written on the diagram giving the reader an overview of the pattern of external inflow of resources, the inside storages in structure, the processes, and the feedback control actions. Heat equivalents are the Calories of heat obtained from each form of energy if converted into heat. Since transfer into heat by definition and by the first law is 100%, heat equivalents are the common denominator of all flows. Even flows of material and information have energy accompanying them.

All inflowing Calories must be accounted for in storages or outflows. If the diagram like that in Fig. 2 is in steady state, inflows equal outflows. A heat equivalence diagram is a "first law diagram". There is generally no controversy in concepts about making a first law diagram, although there is ample room for error in getting the pathways and values correct and comprehensive.

Maximum Power Theories

Heat equivalence measures, or first law measures, provide no information about the potential value of the energy for performing some work function. Second law considerations, however, do. More precisely second law considerations in combination with a time measure of energy flow (which allows energy flow per unit time to be maximized) may, in fact, explain why systems develop certain standard organizational designs. The observed patterns of energy flow and transfer in many kinds of systems seem readily explained by the theory of maximum power. This theory, if general, may make possible the restructuring of science to view systems of many kinds as special cases of a few general patterns. The similarity in the design of systems helps the process of energy analysis, since energy diagrams can be drawn more easily when the basic plans for the shapes and configurations of pathways are suspected in advance.

Apparently first clearly stated by Alfred Lotka in 1922 (3), the maximum power principle states that systems which maximize their flows of energy survive in competition. Among the observed properties of real energy webs, which seem to be explained by this principle, are the characteristic patterns in Fig. 3. Here the potential energy in the source is transformed to a new kind of energy represented by the storage. Some of it is degraded in the process and some is transformed into a higher quality form with new characteristics. Some of this stored higher quality energy is fedback in loops to interact with and amplify the incoming flow of low quality energy from the source. Systems develop chains of these storage-feedback units forming discrete

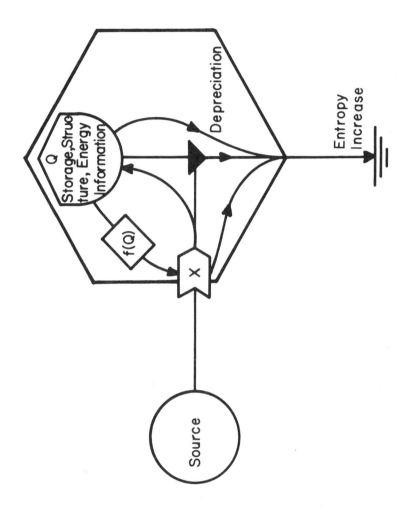

Fig. 3.   Typical sub-unit observed in all systems. Note
          storage, depreciation, feedback, and production
          (transformation work) process.

energy levels. The transformation of energy from low to high quality via webs of storage-feedback units is apparently what allows power to be maximized in the system.

The objective procedures of energy analysis given here do not require acceptance of the generality of the maximum power principle. However, the possibility that all systems can be easily generalized with these energy principles is responsible for some of the excitement developing in this area of science. Details on the maximum power theories are given elsewhere (1).

Characteristic Webs

Figure 4a shows this web of storage-feedback units found in real systems such as those of the biosphere and the economy of humanity. Note that the flow of energy from a primary source simultaneously generates diverging flows that converge back and interact again. Examining any one storage unit on the diagram suggests that several energy inputs are required to sustain that storage. However, tracing pathways back in the web shows that simultaneous diverging and reconverging pathways provide all inputs, each the by-products of the other. For minimum waste the flows can be adjusted so that no one of the necessary interacting pathways is any more limiting to the storage than another.

When the energy from the source or sources on the left diverge, converge, interact, and loop in the characteristic manner shown, potential energy is degraded and dispersed into the heat sink. It is no longer usable for work. As a result, the pathways on the right have relatively few heat equivalents, although their role as feedback controls may be just as important and essential as the flows with larger heat content on the left. The flows and structures on the right require the flows on the left, and vice versa.

Diagram of Cost Equivalents

After an energy web is drawn and the flows of heat equivalents are evaluated, the diagram shows the manner and extent to which the energy flows within the system depend on the sources of energy. Another copy of the energy diagram can be used to write energy costs on all the pathways. This becomes an energy cost diagram. Figure 4b is an example. It is the same as Fig. 4a but evaluated in energy units of equivalent quality rather than in heat equivalents. The energy cost in solar equivalents of each flow is written on the pathway. Since there is only one source for all flow pathways in this example, all pathways have the same numerical cost value. The values on an energy cost diagram are

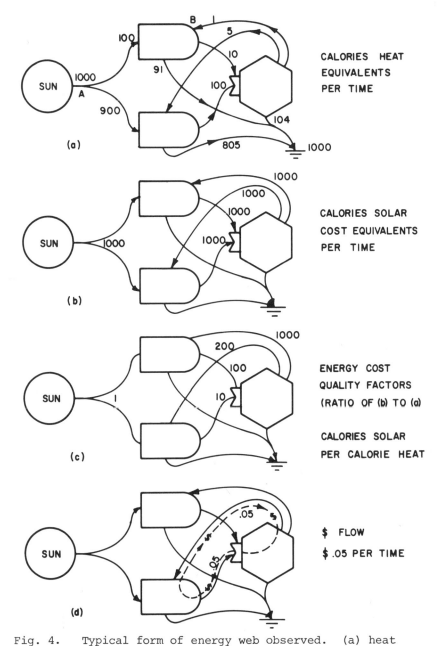

Fig. 4.   Typical form of energy web observed. (a) heat
equivalent numbers included to form a first law diagram;
(b) with solar cost equivalents written on pathways;
(c) with solar energy quality factors written on pathways; these numbers were obtained by dividing those in
Fig. 4a by those in Fig. 4b; (d) dollar flow.

not additive. Pathways diverging from a production process
each have the same cost equivalents. When they reconverge in
an interaction process, the output is not the sum of the con-
verging flows. The cost value is that of the flow originally
responsible for the interacting flows. In this example, the
sun's flow is the cost of all the derived renewable flows.

For several purposes of energy analysis, the equivalent
cost diagram is a basic tool for determining which flows are
important. In it all numbers are expressed in Calorie equivalents of the same type.

In examples where there are two different outside energy
sources, the energy cost equivalents of two interacting flows
may be greater than the cost in cases where all flows are
mutual by products of one source. In that case, observed
energy cost equivalents on the diagram may not be the thermodynamic minimum cost.

Evaluation and Significance of
Energy Quality

The discussion above indicated that the heat equivalent
of an energy flow does not reflect the energy cost required
to sustain the flow. The energy cost of sustaining a flow
or a storage is a measure of its energy quality. Many heat
equivalents are lost to the heat sink when low quality energy
is transformed to high quality energy. The more transformations that occur, the fewer heat equivalents that remain.
But, as we have seen, the high quality energy with few heat
equivalents is required via feedback to maintain the pre-
ceeding transformations. Needed is a means to evaluate this
energy quality at each step. This section introduces a
method for that evaluation and suggests that energy quality
may be a measure of value.

Work and a Scale of Energy
Quality Transformation

Maxwell defined work as energy transformation. Represented by Fig. 4a and observed in systems of all kinds are
chains of energy transformation in which the Calories of heat
equivalence are gradually converted into degraded heat of
low quality while upgrading the remaining energy stepwise into higher and higher quality (Fig. 4). For example there
are food chains like that in the Silver Springs diagram of
Fig. 2. Similar chains occur in the energy transformations
of the human industrial economy, in the chains of energy
transfer in the earth's processes and the chains of bio-
chemical action in cells, etc.

If a system based on one energy source has been maximized for power transformation with the least waste (as

compared to alternative designs), then the ratio of two flows
in a web diagram is the efficiency with which one type of energy flow is transformed into another. For example, in
Fig. 4a the ratio of B to A is 0.1%. The reciprocal is the
number of Calories of one type of energy required to generate
another type. In this case, 1000 Calories of flow A are required to generate 1 Calorie of flow B. This energy quality
ratio is defined as Q.

$$Q= \frac{\text{Calorie flux of type A}}{\text{Calorie flux of type B}} \quad \text{in Calories per Calorie}$$

If the type of energy which is the input is put in the
numerator and the type of energy that results from the transformation is the denominator, then the energy quality ratios
are greater than 1. The energy quality idea is simple in
chains from single sources. It is simply the energy of one
type required to develop energy of another type and is a
cost measure of the relative value of two types of energy.
The ratio of the two flows of energy in heat equivalents is
the energy quality factor for that transformation. It is
hypothesized that there is a minimum energy cost for a transformation at maximum power. That cost represents an inherent
thermodynamic limit below which no improvement can be made.
It is further reasoned that systems that have had a long
period of evolution and survival under competition have approached these thermodynamic limits. Thus it is useful to
develop tables of energy quality factors by evaluating energy analysis diagrams of long established systems. When
there are two sources, energy quality is calculated by expressing one source in quality units of the other type,
using energy quality factors relating the two types of energy as independently determined (4). The complex web of
varied flows that develops is apparently necessary to maximizing each flow. Cost factors can be given in solar equivalents or in units of some other type of energy. Coal
equivalents are often used. An analysis of a system which
transforms the energy of the sun into wood and then into
heat engines, indicated that 2000 Calories of sun are required to produce 1 Calorie of steam, a Q ratio of 2000
Cal/Cal. Do the geological processes which produce coal
from sunlight do better?

Diagram of Energy Quality Ratios

Having drawn a first law diagram and a cost equivalence
diagram, numbers for a diagram of energy quality ratios (Q)
are obtained by taking the ratio of the cost equivalents to
the heat equivalents (as in the example of Fig. 4c). This
diagram shows the solar Calorie cost of each Calorie of

other type of energy flow. As one moves further downstream
from the energy source, the energy quality ratio increases.
Sometimes a table of cost equivalents is used to evaluate the
diagram of energy quality factors which is then used with the
first law diagram to calculate the diagram of cost equivalents.

Cost and Potential Effect

Procedures thus far have shown how to calculate the
energy cost of sustaining some component of a system. But
how can the effect of that component on the rest of the system (via its feedback pathways) be evaluated? In other
words, what is the value of the pathway to the system? The
maximum power theory suggests that the energy cost of a component determines how its feedback flow will interact upstream. For the long selected system, energy costs may have
been minimized and energy amplifier effects are similar. In
that case, energy cost measures energy effect and, therefore,
is a measure of the energy value of the component to the system. In other words, the ultimate potential value of an energy flow is equal to its minimal energy cost, and it may be
safe to assume that systems which have existed for long time
periods have minimized their energy costs. Furthermore,
maximum power theory as well as observed system structures
suggests that the development of a web of energy flows which
produces many kinds of energy at the same time is the most
efficient way to transform energy to higher quality.

For new, developing systems such as some new energy
technologies, energy costs may not be minimized as yet. In
those cases, present energy costs may exceed their effect.
From the maximum power principle, however, it may be postulated that any unit that does not feed back with an amplifier effect that is at least as great as its energy cost may
be a liability and will tend to be eliminated.

When humans manipulate the energy flows in the economy,
they affect the manner in which feedback flows interact with
and amplify the upstream processes. Flows of energy which
have high potential value (because of their high inherent
energy costs) should be saved for uses with high amplifier
effects.

Many new technological mechanisms for energy transformation arranged by man seem simple at first glance. But an
energy diagram of those mechanisms (which forces one to
identify the sources of energy) indicates that large amounts
of high quality energy from a complex web of natural and
economic interactions sustains the new mechanism. The new
mechanisms may use more energy than natural processes. For
example, Kemp (13) analyzed desalination plants and found
that the production of 1 calorie of chemical free energy of

fresh water required 3.1 calories in energy cost expressed as
coal equivalents. This is about 6000 calories of energy cost
in solar equivalents and is higher than natural desalination
by the sun in world weather processes (3215 cal/cal as given
in Table 1).

Paradox of the High Energy Cost of
Flows of Low Calorie Content

Implicit in this discussion is the fact that the energy
cost of maintaining a flow or component increases as the heat
calories that flow contains decrease. It is postulated that
this concept is general because it is a property of all real
energy webs observed. In energy diagrams, such as that given
in Fig. 4, the less a flow at the right seems to involve heat
equivalent energy, the more heat equivalent energy there is
behind it making that flow possible. Flows of valuable
materials, information, human service, etc. seem to be low in
energy whereas the energy flow that makes them possible may
be very large.

Webs of Energy Flow in Nature
and in the Economy

This section applies the concepts developed thus far to
some examples of real world energy webs. By applying the
concepts of energy cost, energy quality, and energy effect
as well as the possibility that systems organize themselves
into webs which maximize power, a great deal of insight into
how real world systems function is possible. First consider
the earth's surface and its biosphere where the energy web
is mainly based on solar energy. Second, consider energy
webs controlled by humans with economic components.

Solar Based Energy Web of the
Biosphere and Earth's Surface

Usually the flows of energy in the biosphere are considered in parts as dictated by such discipline boundaries
as meteorology, oceanography, and geology. But energy flows
across discipline boundaries in the real world. The real
world biosphere system operates as a web with all parts working in unison. Fig. 5 represents an attempt to diagram the
many kinds of energy transformations and feedbacks that take
place in the biosphere as it develops the wind, waves, and
rain and its land cycles, chemical transformations and biological productivities. In the process of diagramming the
biosphere model, many controversial questions were raised.
Before all the pathways can be evaluated with confidence,
some of these questions will need detailed analysis and some

require advances in science. Current calculations in heat
equivalents are given in Table 2 and on Fig. 5 (a first law
diagram). Part of Table 1 was assembled from the ratios
found.

The point is that energy analysis models are one way of
stating hypotheses for further testing. For example, according to older theories, the uplift of land in mountain
building receives energy from the residual temperature gradient between the deep earth and the surface (note the flow
from residual deep heat to continents in Fig. 5). An alternative theory, which emerged as Fig. 5 was being developed, is that there is enough energy from the sun going into
crustal work to drive most of the uplift cycle. Note (Fig.
5) that energy from the sun becomes part of uplift processes through the hydrological cycle, through chemical potential energy deposited in sediments from photosynthesis and
other biosphere activities, and from the heat from radio-
active substances that are concentrated into the surface
cycle by differential photosynthetic, sedimentary, and geo-
thermal activity.

The heat emerging from the earth as potential energy
is about 1.27 calories per square meter per day (12). For
a temperature gradient of 300°C (from 600°C to 300°C over a
depth of 35 km) the Carnot efficiency with which work could
be done is 50%. Such a system, if operating at maximum
power, might do mechanical work with 25% efficiency and produce 0.32 calories per square meter per day as mechanical
work. Figure 5 shows more than this much work in rivers.
The photosynthetic production buried in sediments is large
enough to account for a good part of the emerging heat.

Energy Webs Controlled by Humans

Where pathways in a web are controlled by humans,
money circulates in closed loops and flows as a counter current to the flow of energy (see Figs. 4d and 6). How and
under what circumstances are the money flows and the energy
flows related?

In order to examine the relationship of energy and
money, we consider four cases: the relationship (i) at the
point where energy obtained externally enters a system, (ii)
within a circulating money-energy loop internal to a system,
(iii) in the overall U.S. economic system, and (iv) in circulating money-energy loops at the end of the system web
(the terminal or most down-stream point in the system).

Consider Fig. 6 in examining the point where energy
enters the system. Money (the dotted lines on Fig. 6) circulates around the feedback loops involving humans but not
around the pathways of the environmental systems nor does
it flow out of the system toward the sun or fuels in the

Table 1. Energy cost equivalents.

Type of energy	Table footnote	Solar Calories per Calorie
Solar energy at earth surface	1	1.
Tropical moist air	2	3.3
Winds	3	315.
Gross photosynthesis	4	920.
Coal	5	2027.
Tide	6	3400.
Water World rain chemical free energy	8	3215.
Rain potential over land, 875 m	9	3870.
Potential organized in rivers	10	10,950.
Chemical potential energy over land	11	15,320.
Electricity	12	7200.
Human service in world	13	257,000.
Human service in U.S.A.	14	418,000.
Work of land uplift	15	9.2×10^{11}

Footnotes to Table 1 (continued)

1. One by definition.   2.   Ratio of 4600 to 1400 in Table 2.
3. Ratio of 4600 to 14.6 from Table 2.   4.   Ratio of
4600 Cal/m ^{2}/day to gross production estimate for earth of
5.0 Cal/m^2/day (9).   5.   16,000 x 10^{12} Cal solar energy
estimated to produce 5.65 Calories of electricity in a wood
power plant.   Coal equivalents are 3.6 x electrical (see
footnote 12) equal 22.6 coal equivalents; 14.7 coal equivalents were used in necessary feedback of woods services and
work of collecting wood.   See reference (4).   6.   Energy
analysis of tidal electric plant at La Rance, France,
expressing flows in coal equivalents (CE) in Fig. 7.   Solar
equivalents of coal taken as 2000 Cal/Cal CE as in footnote
number 4.   8.   Ratio of solar energy (4600 Cal) to chemical
free energy in rain (1.43 Cal/m^2/day).   Total rain of world,
520,000 km^3/yr (9); 5.1 x 10 $^{-4}$ Cal/g of rain from footnote
6 in Table 2.   Calculation like that in footnote 6 of Table

2.   9.   Ratio of 4600 Cal solar energy to 1.19 Cal rain
energy overland per m^2 per day, calculated as in footnote 5
of table 2 except continental rain used: 105,000 km^2/yr (9).
10.   Ratio of 4600 to 0.42 in Table 2.   11.   Ratio of 4600
to 0.30 in Table 2.   12.   Ratio of Calories of coal input
to electric power plant to Calories of electrical output.
Input includes coal used as fuel and coal equivalents of
dollars spent on goods, services, and materials.   See
reference (4).   13.   Ratio of 5140 Cal/m^2 /day (4600 plus 540,
the solar equivalent of world fuel consumption) to 0.020
Cal/m^2/day in Table 2.   14.   Ratio of U.S. Solar equivalents
(1.26 x 10^{17} Cal/day) to human Calories (3.0 x 10^{11} Cal/day).
Solar equivalents are sum of U.S. solar energy (solar energy
overland: (1.2 x 10^{6}Cal/m^2/yr) (9.4 x 10^{12} m^2) = 10.6 x
10^8 Cal/yr plus solar equivalent of fuel consumption (2000 x
17.6 x 10^{15} Cal/yr.) Human Calories are product of 2500
Cal/person/day and 120 million people = 3 x 10^{11} Cal/day.
15.   Ratio of 4600 Cal to 5.0 x 10^{-9} Cal in Table 2.

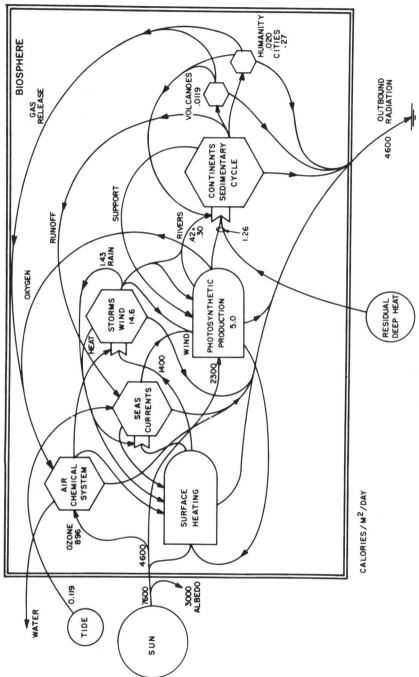

Fig. 5.   Aggregated energy flow model of the main processes
of the earth's surface and the biosphere.   See Table 1
for numbers.

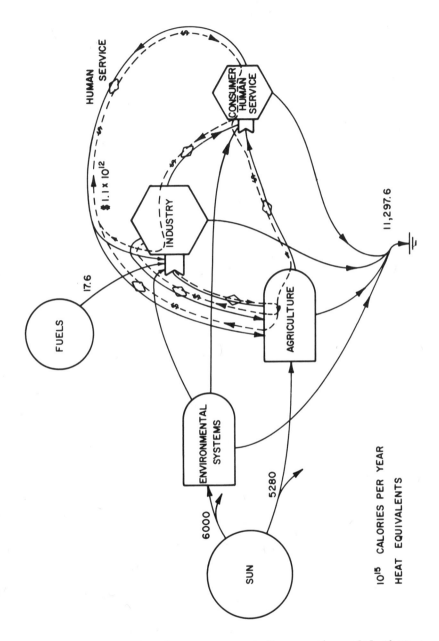

Fig. 6.   Energy flow in an aggregated economic model that
shows the relationship of human service as high quality
feedback in the U.S. Personal income, farm area and
fuel use for 1974 (U.S. Statistical Abstract).

Table 2.   Estimates for environmental energy flows of the biosphere in order of quality.   See Fig. 5.

Type of energy	Table footnote	Heat equivalents Calories/m^2/day
Solar energy not including albedo	1	4600.
Solar energy reaching surface including heat reradiation from sky	1	9000.
Evapotranspirational energy flux	1	1400.
Ozone absorption process	2	896.
Wind and storms	3	14.6
Photosynthetic productivity	4	5.0
Potential energy of rivers against gravity over continents	5	0.42
Potential energy of rain purity compared to sea water over land	6	0.30
Tide	7	0.119
Human labor	8	0.020
Volcanic activity	9	0.0119
World fuel consumption	10	0.265
Gravitational work of land uplift	11	5.0×10^{-9}
Seismic activity	12	2.0×10^{-14}

1.   Sellars reference (10).   2.   13% of insolation, Ryabchikov (9).   3.   Hubbard (8).   4.   reference (9).

5.   River runoff, 37,000 km^3/yr;   average elevation, 875 m (9)

$[(10^2 cm/m) (3.7 \times 10^4 km^3/yr) (10^{15} cm^3 km^2) (875 \text{ km}) (1 \text{ g}/cm^3)$

$(10^3 cm/sec^2) (2.38 \times 10^{-11} Cal/erg)] \div [(365 \text{ days/yr})$

$(5.1 \times 10^{-14} m^2 \text{ area of earth})] \cdot$

Footnotes to Table 2 (continued)

6. Calories free energy per gram of water = RT ln (100/97.5)

$$\frac{(2 \text{ Cal/deg.-mole}) \ (300 \text{ deg.}) \ (0.0154)}{(18 \text{ g/mole}) \ (1000 \text{ g-cal./Cal.})} = 5.1 \times 10^{-4} \text{Cal./g}$$

Continental rain, 109,000 km^3 (9).

$[(109,000 \text{ km}^3) (10^{15} \text{cm}^3/\text{km}^3) (1 \text{ g/cm}^3) (5.1 \times 10^{-4} \text{Cal/g})] \div$

$[(5.1 \times 10^{14} \text{m}^2/\text{earth}) \ (365 \text{ days/yr})] = .30 \text{ Cal./m}^2/\text{yr}.$

7. .0058 watts/m^2(8); unlike other flows, tide is not from sunlight.

8. $\dfrac{(4 \times 10^9 \text{people}) \ (2500 \text{ Cal./person/day})}{(5.1 \times 10^{14} \text{m}^2/\text{earth})} = 0.02 \text{ Cal/m}^2/\text{day}$

9. .00058 watts (8).

10. 1970 3 x U.S. consumption

$$\frac{50 \times 10^{15} \text{Cal/yr}}{(5.1 \times 10^{14} \text{m}^2/\text{earth}) \ (365 \text{ days})} = 0.265$$

11. 3.6 cm uplift per 1000 years (7); 29% of earth surface continental (9)

$[(10^3 \text{cm/sec}^2) \ (3.6 \text{ cm}) \ (.29 \text{ continental}) \ (3.6 \text{ cm}) \ (2 \text{ g/}$

$\text{cm}^3) \ (2.38 \times 10^{-11} \text{Cal/erg}) \ (10^4 \text{ cm}^2/\text{m}^2)] \quad \div$

$[(365 \text{ days/yr}) \ (1000 \text{ yrs})]$

12. $(1500 \times 10^{20} \text{ergs/yr}$ (11)

$$\frac{(1500 \times 10^{20}) \ (2.38 \times 10^{-11} \text{Cal./erg})}{(5.1 \times 10^{14} \text{m}^2/\text{earth}) \ (365 \text{ days/yr})} = 2 \times 10^{-14}.$$

ground. Clearly, the amount of work (energy effect) that
goes with the circulating flow of money depends on those
external inflows from the sun and fuels. But the money flow,
at the point where external energy flows into the system,
buys only the work that is being fed back from the economy
that processes the energy. At that point, money does not
reflect the eventual effect of the external energy. Therefore, the money flow at that point is not proportional to
the amount of energy entering from the external source.

      Second, consider the relationship of money to energy
within a loop internal to the system. How do economic price
mechanisms affect these internal energy flows? By elimina-
ting limiting factors, the price mechanisms of an open market
tend to facilitate the maximum flow of power through the
whole network. For example, when a commodity becomes scarce
and the price rises, more money (and thus more energy) flows
through that pathway from upstream; that is, more money flows
through the pathway in which the shortage occurs. The result
is elimination of the shortage. When a commodity is scarce,
obtaining some of that commodity results in more energy
effect than under non-scarce conditions (because obtaining
the commodity opens a bottleneck of flows). Thus, that
commodity is temporarily more valuable and justifies more
energy cost. It is well established in economics that money
flows into a pathway in response to the marginal effect of
that pathway as a limiting factor. More money flows toward
the commodity that is limiting output than toward any other
commodity involved in producing the output at that time. It
appears then that money flows are proportional to energy
costs when energy costs and energy effects are equal. Fig. 4
represents such a case. However, in the more usual examples
of the present time where the economy is in a transient state
and is heavily subsidized by fossil fuels, some energy flows
are being used with less energy effect than their energy
cost. In these cases, money flow and energy costs are not
proportional. Separate money and energy diagrams identify
such cases. To show the full facts of systems of energy and
economics, a separate diagram of money flow should be included with the first three already mentioned (Fig. 4d).

      Third, consider the case of the U.S. economic system.
The overall money circulation (real GNP) can be related to
the overall rate of energy inflow as a Calorie to dollar
ratio. This ratio changes with time and measures overall
inflation. The ratio of energy inflow to dollar of GNP decreases with inflation. While one can calculate an energy
to dollar ratio where the energy counted is only that of
concentrated fuels, a more meaningful ratio includes all
sources, solar energy as well as fossil fuels. As indicated
by Fig. 6 and others, the money flows depend on solar energy

(as it is processed by the environment) as well as on concentrated fuels. Several questions were raised at the symposium about the possibility of double counting where the
ratio of GNP to total energy flow is used to estimate the
approximate energy contributions of goods, services, labor,
and other inputs to a sector. These questions are addressed
in a note (14).

The final case is that of the money-energy relationship
at the end or termination of the web. In a system like that
of Fig. 6, the high quality pathways at the end of the web
(the far right on the diagram) contain a flow of energy
which is the convergence of most of the energy interactions.
These terminal flows may have nearly the same ratio of energy (in cost equivalents) to dollars as the overall system
does. Given data on the flow of dollars in these terminal
high quality loops, an estimate of the energy flow (in energy cost equivalents) can be obtained by multiplying by the
energy flow/dollar flow ratio for that year.

Among the high quality loops at the high quality end of
a web are the feedbacks of human service. These have very
high energy quality factors and high amplifier control actions at their work interactions. Energy to dollar ratios
are appropriate for estimating the energy cost involved in
these feedbacks. The Energy Quality of a medical doctor's
service may be as high as 4×10^{12} solar Calories per Calorie.

Considerable controversy exists as to what part of the
energy support of humans as consumers is a regular necessary
part of the support of the feedback. Maximum power theory
and experience in analyzing systems suggests caution in dis-
missing as unimportant any part of a working and competing
system. Because of its high quality and thus high energy
cost, human service is the major part of any energy analysis
and cannot be omitted.

Evaluating Alternatives

After energy diagrams are prepared and energy quality
factors estimated, special calculations can be made to suggest which features of a system or proposed system are ener-
getically important. Examples of such calculations are
given in this section for some cases of special interest in
energy-environmental policymaking.

Evaluating Net Energy

Net energy is the difference between the yield of energy
and the feedback required in a process, where both flows are
expressed in Calorie equivalents of the same type. A net
energy calculation is made to evaluate a single source to

La Rance, France

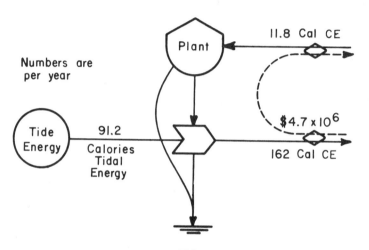

Energy Quality Factor = $\dfrac{150}{91}$ = 1.7 Cal CE/Cal. Tide

Yield Ratio = $\dfrac{162}{11.8}$ = 13.7

Net Energy = 162 - 11.8 = 150 x 10^{10} Cal CE

Fig. 7.   Example of net energy evaluation of a single
source. Tidal energy converted to electricity. Both
electricity and feedbacks of goods and services are
converted to cost equivalents of the same type (fossil
fuels as used in heat engines abbreviated FFE).

determine how important its contribution is. Figure 7 is an example. As in procedures previously described, heat equivalents of the flows are determined first. Then, using tables of quality factors, solar or coal cost equivalents are estimated and written on the diagram. The difference between yield and feedback is the net energy (Y-F).

To interpret the importance of the source to the economy, the ratio of the yield to feedback is calculated. High ratios mean that the source can support the development of more activity in the economy downstream to the right. When the yield ratio is small, there is little energy to support activity other than that which supplies the necessary feedback. A system with only one source which has developed a steady state has no net energy since it feeds back energy of equal cost to that delivered (as illustrated by Fig. 4b). Where there are several sources and/or where there is growth with feedbacks not yet fully developed, analysis of a single source (as shown in Fig. 8) can indicate the role of that source in supporting more economic development. A whole system which is in steady state has no net energy; it feeds all of its work from net energy sources back to amplify interactions, subsidizing other sources, and maximizing power (as illustrated by Fig. 6).

The U.S. is running now on many sources with yield ratios of about 6 units yield for 1 fed back.* Sources with higher yield ratios than this are good primary sources and contribute more to the economy. Sources with a lower ratio are being partially sibsidized by the main economy, since they yield less per unit received back than their competitors.

As was indicated earlier the amount of circulating money associated with the production of an energy source does not indicate the energy contribution of that external source. It only affects the overall energy to dollar ratio later. A source need not be a good one (competitive) or have net energy to be economic.

Evaluating Secondary Sources

A secondary source is one that does not yield net energy although it does bring in additional energy to the system

---

*In calculating the net energy and yield ratios of primary energy sources using the method described here, the energy costs include those associated with concentrated fuels, labor, and solar energy as it is processed by the environment. All of these are necessary inputs and are present in the feedback loops which allow the source to be develop. All must be evaluated in equivalent energy cost units prior to summing.

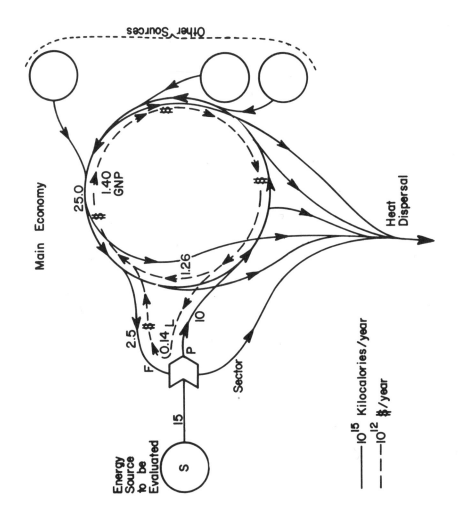

Fig. 8.   An example of evaluation of an economic sector and
source.

from the outside. A secondary source receives more energy
in feedback than it draws from the environment, where both
are expressed in Calorie cost equivalents of the same type.
However, secondary sources are a major, necessary part of
systems that have an excess of high quality energy from one
or more primary sources. High quality energy does not
generate effects commensurate with its energy costs unless
it can interact with large quantities of low quality energy
such as sunlight. For example, energy in rivers and fossil
fuels must be used in interaction with landscapes and solar
energy to generate as much work as these sources cost. The
more the high quality energy can be spread out to interact
with the solar energy the greater amplifier action it may
have. Examples are irrigation, tourism, forestry, and
fisheries. All of these depend on high quality fossil fuel
sources which feedback, interact, and amplify the solar
energy required for crop production, forestry, fisheries,
and tourism. But as these systems are now operated, solar
energy is a secondary source and the high quality fossil
fuels are their primary energy source.

Evaluation of the secondary source interaction with
high quality feedback is done in the example given in Fig. 8.
Heat equivalents are evaluated first. Then cost equivalents
of the same quality are evaluated. Then an investment ratio
is calculated. The investment ratio is the ratio of feedback to the flow of new resource, where both are expressed in
Calories of the same quality. A source is competitive when
high quantities of new external energy are brought in per
unit of feedback energy invested to make the process possible. In the U.S. as a whole, a usual ratio of feedback to
inflow is 2.5 to 1 (both in energy units of the same quality), 2.5 Calories of energy invested via feedback for each
1 Calorie that investment brings in externally. Ratios lower than this are economic; ratios higher than this tend to
be less competitive.

Evaluating Consumer Feedbacks

Some of the higher quality feedback loops of systems,
such as human medical and governmental service, feedback
their work with little direct interaction with external
energy sources. Their contribution to maximizing power in
the system is in providing special mechanisms, materials,
parts, controls, and information. Evaluating their contribution involves comparison of their energy cost with their
energy effect. Energy costs can be obtained from the basic
energy diagrams showing the energy flows required to develop feedback. The effect, however, can be determined
only by disconnecting the pathway and observing the energy
flows with and without the feedback interaction. Often

these numbers are found by comparing similar systems which
differ in having the concerned pathway. Often simulation
models are used. This concept of consumer feedback with
consequent amplifier action on the whole system can be illustrated by three examples, one involving no humans and two
where humans are essential.

A tropical forest plantation of Cadam trees in Puerto
Rico has a productive net yield of photosynthesis 20 g/m^2/
day (80 Calories/m^2/day wood equivalents) as a monoculture
without many consumers (5). In contrast, a fully developed
ecosystem nearby (with fully developed consumers feeding
back in an organized manner) showed an increase in this basic primary production. An increase of 7 g/m^2/day (28
Calories/m^2/day), most of which was used by the consumers
without any net energy, was measured. The system with consumers contained more energy flow (power) than the same
system without consumers. Most of the web of producer-con-
sumer interaction was required to maximize power.

In systems involving human consumers, many think of
human consumption as the terminal purpose of an economy. In
contrast, human consumers really act as units which feedback
services necessary for maximizing power under competition.
Agriculture and space heating provide two examples.

Only in primitive subsistence agriculture was crop production a primary energy source that yielded net energy. In
subsistence agriculture, yield ratios are about 2 to 1. By
the time human activities are coupled back into the system,
the yield ratio is closer to 1 to 1. Most industrial agriculture now receives more energy (in the form of fossil
fuels) back from the economy than it yields (all energies
measured in cost equivalents). Thus industrial agriculture
is now a secondary source of energy. It is characterized by
ratios of feedback to inflow energy of 2-10 to 1 (yield
ratios of 0.1 - 0.5 to 1). When agriculture (or other similar solar technologies) are carried out in tiny areas such
as greenhouses, ratios of feedback energy to inflow energy
are very high, 1000 to 1 or more (both in Calorie equivalents of the same quality), or yield ratios of 0.001 to 1 or
less. Since they take far more energy from the economy
than they contribute, such operations are not sources of
energy. Rather, these operations are consumer devices that
use solar energy to aid the flow of some other kind of energy. For example, greenhouse vegetables could be necessary
for the health of human beings on a desert island; the amplifier action would be that of the health differential and
the energy cost would be justified because of its effect on
human health. The energy effect is to increase the power
flow of the entire system (because the human population is
healthier and can interact and do more work in other parts
of the system).

Neither a gas water heater nor a solar water heater yields net energy. A gas water heater takes 11 Calories to generate 1 Calorie of hot water. An evaluation of solar water heaters as an energy conservation action (in comparison to natural gas heaters) suggests a savings ratio of 4 Calories per Calorie; the system does still not yield net energy. However, space heating is clearly required for human productivity. It should be viewed as a consumer device which is energy costly, but which is also energy effective via all of the feedback pathways involving human productivity.

In summary, excess energy goes to consumers who feedback with an amplifier effect and make the whole system more effective. Undoubtedly in times of expanding energy, a system, which is already ahead of others in competition for power, generates net energy that goes to consumers but does not immediately feedback to amplify some other flow in the system. The maximum power theory suggests that such unlooped consumer flows are fairly random, but are creative, and after later selection, effective feedback interactions develop. As energy excess decreases and growth slows, those feedbacks with greatest effect will survive; unlooped consumer flows will not. In order to plan for times of decreasing net energy, it is important that we begin evaluating the energy cost and energy effect of the multitude of consumer feedback loops existing in our economy.

Evaluating Energy Conservation Alternatives

Measures proposed to conserve energy can be evaluated on a Calorie per Calorie basis. The feedback of conservation service such as providing housing insulation or improving car efficiencies can be evaluated in Calorie cost. Calorie savings can then be compared to the Calories fedback in the savings effort (where both are expressed in energy equivalents of the same quality). If the ratio of savings to feedback is greater than one there is a net energy contribution. The feedback is usually one of high quality goods and services, and data are usually expressed in dollars. The U.S. Federal Energy Administration has sometimes used the ratio of dollars spent to energy saved. This ratio is about the same as the ratio of energy spent to energy saved, since feedback of high quality goods and services can be evaluated with an energy to dollar ratio.

Evaluating Environmental Impact

The use of energy diagrams and energy analysis for evaluating environmental impact has led to some exciting, if

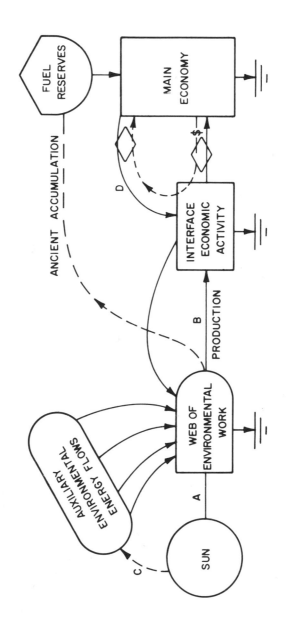

Fig. 9.   Summary of energy flows of the environment
attracting additional energies of investment and economic development.   Cost equivalents are evaluated at
A or B and related to actual or potential energy flows
attracted at D.

controversial, insights into the appropriate use of environmental control technologies. Environmental control techniques are energy costly. The maximum power principle suggests that their energy effect in maintaining human health (e.g. a flue gas desulfurization system) and in maintaining environmental flows or fisheries (e.g. tertiary sewage treatment plants and cooling towers) ought to be at least equal to their energy cost and should involve external solar energy where possible. The investment ratios of these systems (e.g. the ratio of energy invested in a tertiary sewage treatment plant to the energy flow that investment involves in solar energy interaction) ought to be as low as possible. Our analyses at the University of Florida indicate that some advanced technologies have very high investment ratios. In these cases, the energy flow in the environment that is maintained or even amplified by the technology is too small to justify so much economic investment. Technologies with high investment ratios are poor users of the conservation dollar.

A better fit of humanity and nature is obtained by coupling the wastes of the economic system to the natural systems through interface ecosystems which can make more use of solar energy. Fig. 9 shows a general format for evaluating such energy interactions with the environment. An example is the recycling of treated sewage into cypress swamps as was carried out in our Florida experiments (Fig. 10). Compare the investment ratios of two alternatives for handling secondary sewage. A tertiary sewage treatment plant might be invoked to remove the nutrients from the effluent prior to its release into a river. The investment ratio for that alternative is 100 to 1 or more. At least 100 Calories of energy are invested in the treatment plant for each Calorie of productivity in the coastal zone involved in the process (all Calories equivalent in quality). The alternative evaluated in the Florida experiment called for cycling the secondary treated sewage directly into a cypress swamp. The wastes were absorbed or transformed and valuable wood growth accelerated. The energy investment in the system (D in Fig. 9) was 11.5 x 10^6 Calories (coal cost equivalents) per year per acre and represented mainly the cost of pipes and pumps. The energy flow from the swamp (expressed in coal cost equivalents) was 3 x 10^6 Calories per acre per year (B in Fig. 9). The investment ratio (the ratio of D to B in Fig. 9) is 3.8 to 1, a vast improvement over the 100 to 1 ratio involved in a tertiary sewage treatment plant.

Furthermore, the mining and manufacturing processes required to assemble raw materials into a treatment plant depend themselves on environmental energy flows. We have seen over and over again through these energy diagrams that the

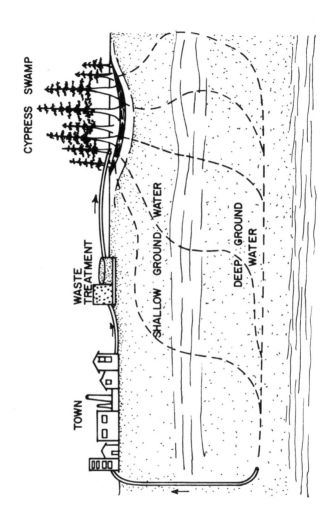

Fig. 10. Example of using a cypress swamp as an interface ecosystem to recycle wastes and maintain a high ratio of useful solar energy to purchased goods and services from the economy (6).

economic processes with which money is associated rarely
take place in the absence of environmental processes (based
on solar energy). The economic processes both interact with
and depend on the environmental ones (e.g. manufacturing
depends on the wind to dilute and disperse its air pollutants). The processes of mining and manufacturing utilize
and load the cleansing capacities of these environmental
flows. For 100 units of energy invested in the tertiary
sewage treatment plant about one third is environmental
loading elsewhere. (The U.S. energy budget matches 2.5 coal
equivalents of fuel energy with one coal equivalent of renewable environmental energy). In the case of the treatment
plant, utilization of 33 units (100 x 1/3) is more than the
1.3 units (3.8 x 1/3) required for the recycling system. In
addition to being poor investments, the distinct possibility
exists that advanced environmental control technologies
actually cause more environmental degradation than they alleviate.

Summary

Energy analysis is the basic science of energetics of
open systems, which considers laws and principles by which
energy flow generates designs of structure and process. A
language of energy symbol diagrams helps develop models and
organize data for analysis and synthesis. Understanding the
contributions of external energy sources and internal mechanisms is aided by preparing diagrams: (i) a first law diagram of heat equivalent flows, (ii) a diagram with energy
costs expressed in Calories of the same quality, (iii) a
diagram with energy quality factors as related to sunlight
or coal and (iv) a diagram with money flows. Energy analysis studies are generating new concepts of energetics, systems organization, power spectra, and the energy basis of
economics.
Practical application of energy analysis includes calculations of net energy to evaluate primary sources, calculations of an energy investment ratio to evaluate secondary sources, calculation of energy savings ratios to evaluate energy conservation ideas, and calculation of energy
effectiveness ratios to evaluate which consumer roles are
competitive.
Because of its generality, energy analysis may be useful as a point of departure in general education of students
learning the unity of the world system of humanity, economics, and environment.

References and Notes

1. H.T. Odum, Environment, Power and Society (Wiley-Interscience, New York, New York, 1971); H.T. Odum and E.C. Odum,
Energy Basis for Man and Nature (McGraw-Hill, New York,
1976); H.T. Odum, in Systems Analysis and Simulation in
Ecology, B. Patten, Ed. (Academic Press, 1972), Vol. 3, p.
139-211; H.T. Odum, in Report of President's Science Advisory Committee Panel on World Food Supply, (White House,
Washington, D.C., 1967), Vol. 3.
2. H.T. Odum, Ecological Monographs 27, 55 (1955).
3. A. Lotka, Proc. Natl. Acad. Sci. 8, 147 (1922).
4. H.T. Odum, C. Kylstra, J. Alexander, N. Sipe, P. Lem et
al, in Hearings before Subcommittee on Energy and Power-
Middle Long Term Energy Policies and Alternatives, (House
Committee on Interstate and Foreign Commerce, Washington,
D.C., 1976), Serial 94-63.
5. H.T. Odum and R.F. Pigeon, A Tropical Rainforest (U.S.
Atomic Energy Commission Division of Tech. Information, 1970)
TID-24270 (RRNC-138).
6. H.T. Odum, K.C. Ewel, W.J. Mitsch, and J.W. Ordway,
Recycling Treated Sewage through Cypress Wetlands in Florida
(Center for Wetlands, Univ. of Fla., Gainesville, 1975),
Occasional Publication No. 1.
7. S. Judson, Am. Sci. 56, 356 (1968).
8. M.K. Hubbard, Sci. Am. 224, 60 (1971).
9. A. Ryabchiko, The Changing Face of the Earth (Progress
Publ., Moscow, Russia, 1975).
10. W.D. Sellars, Physical Climatology (U. of Chicago Press,
Chicago, Illinois, 1965).
11. D.A. Walker, Science 193, 886 (1976).
12. P.J. Wyllie, The Dynamic Earth (John Wiley, New York,
New York, 1971).
13. W.M.Kemp, Ph.D. Dissertation, University of Florida
(1976).
14. Double Counting Questions: Figure 8 shows the flows of
energy and money in the general economy. It includes free
environmental energies and external fuels from deposits
which are also free since their external pathway is not
accompanied by money. All Calories must be in cost
equivalents (not heat equivalents). The rough proportion is
used to estimate the energy feedback (F):

$$\frac{\text{Energy feedback (F)}}{\text{Total energy (T) including environmental inputs}} = \frac{\text{Money flow in Loop (L)}}{\text{Total GNP}}$$

In the example shown this is:

$$\frac{F}{25 \times 10^{15} kcal/yr} = \frac{0.14 \times 10^{10} \$/yr}{1.4 \times 10^{12} \$/yr}$$

$$F = 2.5 \times 10^{15} Kcal/yr$$

Because some of the energy of sources goes into the economy and back to the sector as goods and services, one must correct for double counting for some purposes. When diagrammed with energy circuit and money flows as shown in Figure 8 there is no question about what is meant and no question about what is the correct answer to the net energy questions.

If the question is: how much of the energy of the main economy is feeding back with feedback F, the answer is 2.5×10^{15} Kcal per year of which 40% was originally from the source S, since source S with 10×10^{15} Calories is 40 percent of the total of 25×10^{15}. If the question is what is the net energy contribution of source S., then one subtracts F from P. In the example $10 - 2.5 = 7.5 \times 10^{15}$ Kcal net energy in fossil fuel equivalents. The yield ratio P/F is 10/2.5 or 4/1. In this example the sector is a net producer supporting other sectors.

Suppose the question asked is, "What are the ultimate energy sources for the sector?" In Figure 8, to obtain the total energy basis of the sector one should add the inflow from source (S) to 60% of the feedback (F), since this is the amount of F that is from entirely different sources.

15. The work was supported by the Energy Research and Development Administration through contract E-(40-1)-4398 with the University of Florida and by the National Cooperative Highway Research Program of the National Research Council through a contract with Cornell University.

Richard H. Williamson is assistant director for systems analysis, Office of the Assistant Administrator for Planning, Analysis and Evaluation, ERDA. Among his specific concerns are economic growth and development, the economics of nuc-lear power, and "net energy analysis." He is the United States delegate and chairman pro-tem, Steering Group in Energy R&D Strategy, International Energy Agency, Paris, France.

5

Energy Analysis and Energy RD&D: Planning and Decisionmaking

Richard H. Williamson

In response to the requirement for net energy analysis, the Energy Research and Development Administration (ERDA) set forth some immediate objectives. These objectives were: to obtain an initial energy analysis for each of the generic technologies that were under consideration in order to determine if there were "net energy losers" among them, to improve the definitional and computational aspects of net energy analysis, and to establish some uniform guidelines to be used throughout the agency so that there would be some comparability among energy analysis calculations. We are nearing completion of these tasks and have reached a decision point that, perhaps, will guide the direction of future work.

The purpose of this paper is: (i) to present a status report on what ERDA has done in the way of net energy analyses, (ii) to summarize the energy analysis procedures and guidelines used within ERDA, (iii) to present a few insights and findings from ERDA's current work, and (iv) to discuss how ERDA is currently progressing towards the use of net energy analysis in decision-making.

Energy Analyses Performed For ERDA

Before the ERDA organization was established, energy analyses were performed primarily by individuals examining individual technologies using very different methodologies. Each reported conflicting types of results. The early emphasis in these studies was placed on supply technologies. In order to obtain a relatively uniform analysis of a large number of supply technologies, the Department of Interior, before the formation of ERDA, funded two studies; one by the Colorado Energy Research Institute (1) and one by Develop-

This paper was prepared as a part of the author's duties as assistant director for Systems Analysis, Office of the Assistant Administrator for Planning, Analysis & Evaluation, ERDA.

ment Sciences Inc. (2). These studies came under the
direction of ERDA personnel, when the Department of Interior
people were transferred into ERDA. The two studies, both
published in early 1976, provided the first comprehensive
evaluation of approximately twenty different energy supply
technologies; and surprisingly enough, the findings of the
two studies were in general agreement.

Recognizing that energy analysis should also be applied
to end use energy technologies, and that energy analysis
should involve the complete energy trajectory from resource
extraction to the point of end use, ERDA funded Development
Sciences Inc. to do additional work. This study entitled
Application of Net Energy Analysis to Consumer Technologies
has just been released by ERDA (3). It contains an analysis
of ten end use technologies covering representative technologies in the residential and commercial sector, industrial
sector, and automotive sector. It also contains about
twenty-five complete energy trajectories which combine the
results from the earlier supply technology study with the end
use technology work that was completed as part of this study.
The study includes most of the elements researchers have
recommended and provides the first energy analysis of the
entire trajectory. With this study one can compare any type
of energy system.

In order to complete the initial evaluation of supply
technologies, we later funded the Institute for Energy
Analysis at Oak Ridge to analyze five additional supply type
technologies that had not been covered by earlier studies.
A draft of that study has been completed, and we hope to release that report within the next couple of months.

Both the Institute for Energy Analysis and the Center
for Advanced Computation at the University of Illinois, have
been funded to evaluate existing energy analysis methodology
and to propose methodological improvements. The Institute
for Energy Analysis will publish a monograph later this summer on Net Energy Methodology.

In addition, the Institute for Energy Analysis has been
working on some draft guidelines to be used throughout ERDA.
ERDA has a working draft of these guidelines and will continue to improve them and eventually put them into effect at
ERDA. However, even after these guidelines are completed,
they will remain flexible so that changes or improvements
in the net energy methodology can be incorporated and
accommodated.

Our future work includes another study with Development
Sciences Inc. (the group who has done analyses of complete
trajectories using the complete computational process). That
study will provide ERDA an evaluation of how net energy may
be used in different decision processes as well as an evaluation of the relative importance it should be given in

conjunction with other decision criteria. This report will
be due sometime late this summer. This study, then,
represents an attempt to get out of the computational mode,
in which we now have a first pass-through analysis of many
technologies and systems, and to get into the more serious
mode of the role of energy analysis in our decision-making
process.

Finally, we are in the review stages of a project proposed by Energy Policy Studies, Inc. that will attempt to do
a comparative energy and an economic analysis of one particular technology. Factors that cause divergences between the
results of the two kinds of analyses will be identified and
their affect quantified.

These are the kind of activities that ERDA has pursued.
Many of the activities are coming to a close very soon, and
we think we have achieved our immediate objectives. Now
ERDA is examining longer run objectives and will soon decide
which to undertake.

Procedures

Several procedural aspects of ERDA's approach, I believe,
need clarification. Current procedures divide energy
analysis into two parts. The first part includes analyses of
resource use; the second part includes analyses of direct and
indirect energy inputs to a process. The resource use
analysis represents an evaluation of the physical losses or
the efficiency of the process from resource extraction
through the entire trajectory to end use. For example, the
amount of resource left in the ground when the extraction
process takes place is a resource loss in ERDA's net energy
terms. Similarly, the loss at a central station power plant
due to the thermal efficiency of the power plant is a resource loss. Losses in any refining steps, or other parts of
the fuel cycle, would be resource losses. We keep that
resource flow of energy separate from the second part of the
analysis which includes the direct and indirect energy inputs,
sometimes referred to as energy subsidies. This is the
energy required to execute a process in one step of a
trajectory. It includes both the direct energy or fuels
necessary to operate a particular step in the fuel process;
it also includes indirect energy, similar to that described
by Clark Bullard in this book. Both the resource loss and
inputs are computed per 1,000 Btu's of energy output; that
is, they are typically normalized to 1,000 Btu's output and
presented in that form. The resource flow data are kept
separate from the energy subsidy data; the two are not added
together. The two kinds of data provide different information.

In addition, the current process used at ERDA combines

92 Richard H. Williamson

input-output analysis for calculating indirect energy with
process analysis for calculating the direct energy flows.
The energy inputs associated with the environment, with
labor, and with the infrastructure of the economy, which
Dr. Odum includes in his analyses, have not been included.
The published calculations on the environmental energy
subsidy, show that the values are rather insignificant when
compared to the other energy inputs. The question is whether
it is fruitful to pursue, at an aggressive level, calculating
the environmental energy subsidy when it is several orders of
magnitude below the other energy inputs. While the issues
that Dr. Odum is making with regard to the environment are
very important, I really question whether net energy is the
arena in which they should be made.

The reason that the energy embodied in labor and in the
infrastructure of the economy has not been analyzed is
because the conceptual grounds for including it is not well
established at this time; there is a great deal of controversy among researchers as to where that energy belongs.
Until the conceptual problem is cleared up, we do not feel
that it is worthwhile to include either labor energy or
infrastructure energy in our findings as a government agency.
In addition, the computational procedures related to calculating that energy input are far from being developed.

Findings, Insights, and Observations

What are the insights that have been gained at ERDA from
the net energy studies carried out so far? First, there are
no energy processes that have been determined to require a
greater energy subsidy than the energy they produce. That
is, there are no net energy losers. Nor have the calculations shown that any technologies are any where near being
losers.

Secondly, ERDA's studies indicate that one should not
use a single index to rank technologies on a net energy basis.
The reason is that the definition of the system's boundary
for computing an index can be shifted all over the place to
yield almost any index desired. In addition, there remain
the problems of different quality,and of mixing resource loss
with energy subsidy (a procedure used in some analyses and
one that simply masks the important problems of the energy
analysis). Many index definitions proposed seem plausible,
all seem to be rather logical and all generate rather widely
different values. For these reasons, I believe the index
value masks the information to be learned from energy analysis. It is more important to keep track of the flows of
energy and the various points in the system where energy is
consumed. This disaggregated information will be more
helpful in aiding decision processes than will a single index.

Some analysts have asserted that net energy estimates will not change with changing dollar values. ERDA's studies indicate that that is incorrect. Particularly when assessing the entire trajectory, one readily finds that substitutions of processes will occur in response to changes in dollar values, or one finds that substitutions of production factor inputs change as dollar values change. A net energy value estimate is not an immutable value. It will change with changes in processes and with substitutions that occur in response to other factors.

The concept of an energy theory of value whereby all decisions are based solely on the energy efficiency of processes is an extreme concept that should be quickly discarded. Energy is but one factor of production and one factor of decisions. Decisions must balance net energy against many other factors.

Some analysts have suggested that a new technology could be implemented at such a rapid rate that the energy inputs required to construct the increasing numbers of new facilities would always be greater than the amount of energy the operating plants could produce. This fact has been mathematically proven. However, such rapid rates have little application in the real world where rates of growth follow more normal balanced processes.

With the exception of additional work on system boundary problems and system boundary definitions, improvements in current methods for computing net energy values will not change the current estimates sufficiently to justify undertaking new studies, until such time as the role of net energy in decision processes has been better defined. These are some of the findings which result from my evaluation of the net energy work that has been carried out so far.

I would like to make one other observation on the role of net energy and energy analysis. Many analysts have eliminated a concept that I think is important, the concept of opportunity cost. In economics, the opportunity cost is the cost of the next best alternative use of the resource. In other words, the concept of opportunity costs asks the question: what else could you do with the resource besides use it in the current process under consideration? While opportunity cost is an economic concept, I believe its application to net energy analysis is quite appropriate. When we consider making decisions on the basis of energy analyses, we should also ask what else could be done with the energy, or what else could be done with the resource. For example, consider a straightforward case: the opportunity of using natural gas in a central station power plant versus using it directly to heat homes. All of the net energy analyses have shown that it is preferable to use it directly to heat homes rather than run it through a power plant before

heating homes. Here is an opportunity use of natural gas;
one ought to make decisions based on it. On the other hand,
consider the use of uranium. If uranium is not used in
nuclear power plants, what is its opportunity use (besides
making bombs, which has nothing to do with the energy
problem)? The fact that there is a resource loss involved in
making electricity from uranium should be assessed against
whatever other opportunity you have to use the uranium. (Let
me make clear that I am not making all the other arguments
pro or con on nuclear power but am considering net energy
only). Only if the net energy subsidy or the energy inputs
to build and operate the power plant were substantially
greater than any other power plant facility would one reject
nuclear power on net energy grounds. One would not reject it
on the basis of efficiency losses. Shale oil provides a
similar example. About three years ago, an alleged study by
an anonymous author indicated that it took eleven units of
energy to produce ten units of shale oil. We have never been
able to determine who actually produced that report, but it
gained widespread attention. With regard to opportunity
costs, suppose it took eleven units of natural gas to produce
the ten units of shale oil. That is definitely a net energy
loser because there are other opportunities to use the natural
gas. But what if it took two units of natural gas and nine
units of shale in situ. Is that an energy loser in light of
opportunity costs? What else can one do with the shale oil?
Energy analysis techniques ought to factor opportunity costs
into the analyses in a better manner than has been done to
this point.

Energy Analysis in Decisions at ERDA

I will conclude by discussing where I think the best
prospects for net energy use in the R + D decision-making
process at ERDA lie. The first application lies in the area
Dr. Ross talked about in his paper. Energy analysis can
help focus R + D on the opportunities for improving the process efficiencies of new technologies. Whether this is a
first law or a second law analysis is not really important.
What we have now are calculations of the net energy of a
process. What is still missing is some definition of a
reasonable or practical net energy value that one might expect to achieve through some innovation. If there is some
indication that substantial improvement in the net energy of
a process is possible, that process becomes a target of
opportunity for research and development. But if there is
very little opportunity for improving that process on net
energy grounds, that process is not much of an opportunity
target for research and development. In this area, energy

analysis can play an important role in R + D decision making.

Secondly, energy analysis may play a role in determining when a new technology might best be introduced into the marketplace. If an energy process has a certain efficiency now and could be introduced in the market (presuming that the economics were favorable), one might decide to delay the introduction of that process until some time in the future when a better efficiency could be achieved. But only if one had in mind some way to achieve that better efficiency could such an argument be made.

Thirdly, energy analysis may play a role in augmenting economic analysis by acting as a check on price resource pressures. There is the possibility that the economics of a process may diverge from the results of an energy analysis of the same process for reasons that are not expected. I personally expect them to diverge in a number of cases for very good and logical reasons. But, if the results of an energy and economic analysis diverge for reasons we do not expect, some important insights and information would be provided.

Other uses for energy analysis are uncertain at the moment; ERDA is investigating about another half dozen different types of potential uses of energy analysis as they relate to our decision-making process. However, I must conclude by saying that, based on todays knowledge on the subject, net energy analysis is playing a fairly limited role in terms of energy RD + D decision-making. I think the reason for this is that its potential for improving the decision-making process has yet to be demonstrated. I am not saying that it can not be demonstrated, but that it has not yet been demonstrated. The how and where to include energy analysis in the decision process has not yet been figured out, and its superiority to other decision-making criteria has not been proven. ERDA will continue to pursue energy analysis research and will continue to investigate how it might factor into our decision-making process, with the hope that energy analysis will improve that decision process.

References

1. Colorado Energy Research Institute, <u>Net Energy Analysis:</u>
   <u>An Energy Balance Study of Fossil Fuel Resources</u>
   (Colorado Energy Research Institute, Golden, Colorado,
   1976).

2. Development Sciences Inc., <u>A Study to Develop Energy</u>
   <u>Estimates of Merit for Selected Fuel Technologies</u>
   (Development Sciences, Inc., East Sandwich, Mass.,1976).

3. Development Sciences Inc., <u>Application of Net Energy</u>
   <u>Analysis to Consumer Technologies</u> (Energy Research and
   Development Administration, Washington D.C., ERDA 77-14,
   1977).

Summary

Martha W. Gilliland

As indicated in the introduction to this volume, the
five chapters reflect a spectrum of thinking as to the policy
applications of energy analysis. In so doing these chapters
identify major elements of the present controversy over the
usefulness of energy analysis to policy-making. This summary seeks to clarify the theoretical and methodological
differences reflected in the different approaches and to
summarize the potential policy applications of each type of
analysis.

Three different conceptions of energy analysis are
presented in the papers by Bullard, Ross, and Odum. Alternatively, Gilliland discusses the potential policy applications
of each conception and clarifies the kinds of information
each provides. She then compares each with conventional
economic analysis. In the final chapter, Williamson
discusses both the manner in which the Energy Research and
Development Administration (ERDA) uses energy analysis and
the kinds of energy analysis research projects ERDA has
supported.

Much of the confusion surrounding energy analysis has
arisen because analysts calculate energy costs differently;
that is, they count different kinds of energy and they perform the accounting differently. Initially this summary
examines the differences in what is counted and how the
counting is done by those using the different techniques.
The second section summarizes the policy applications of
each technique. It is not my intent to identify one
technique as better than another. Quite to the contrary,
each addresses different policy questions and provides
different kinds of useful information.

What Is Counted And How It Is Counted

There appear to be three categories of energy that are
required by man-technological systems: concentrated fuels

(e.g. coal, oil, gas, uranium), labor, and environmental
energy. It is universally accepted that fuels are energy.
It is widely recognized that labor is a form of energy and
that it can often substitute for fuels, but attempts to give
labor an energy value are not widely accepted. Similarly,
while the dependence of technology on environmental energy
has been widely recognized only in the contemporary period,
most people now agree that it would be impossible to carry
out industrial or technological processes or any human
activity in the absence of, for example, the hydrological
cycle, atmospheric circulation, or photosynthetic processes.
Few agree, however, that we know how to quantify that dependence in energy units, measure its actual value to man or to
technological systems, or substitute it for concentrated
fuels.

The techniques presented and applied in Bullard's and
Ross's papers are designed to evaluate the energy in concentrated fuels only. Bullard's paper does add data on labor
costs and employment impacts, but the units used are jobs
not energy. In contrast, the techniques and theory presented
by Odum are designed to evaluate the energy contribution from
concentrated fuels, labor, and the environment. Theory,
methods, and system boundaries are fundamentally different
among the techniques. We emphasize, however, that each
technique is designed to answer different questions.

The following discussion summarizes the manner in which
concentrated fuels are evaluated using the approaches presented by Bullard and Ross. Following that the manner in
which concentrated fuels, labor, and environmental energy are
evaluated using Odum's approach is discussed. A central concern will be with the theoretical bases of the various
approaches.

Concentrated Fuels

Energy analysis is derived from the laws of thermodynamics, but the techniques discussed by Bullard and by Ross
vary in the extent to which those laws are used. Specifically, the analytical approach presented by Bullard measures
energy flows in terms of their energy content (enthalpy).
Such measures are known as first law measures, after the
first law of thermodynamics. The approach presented by Ross
incorporates second law considerations into energy measures,
measuring energy flows in terms of the available work they
contain. Such an analysis, of course, requires knowledge of
energy content also; that is, it also requires first law
knowledge.

In the first law case, the physical commodities that are
required by a process or policy are evaluated in terms of
their energy content. Physical commodities include fuels,

goods, and services. Thus, the energy content of the fuels
used directly by a process or required by a policy is simply
the enthalpy or heat of combustion of the fuel; the energy
content of the materials used or required is the fuel used to
manufacture (and transport) the materials. Similarly, the
energy content of services is the fuel required to provide
them. After evaluating all the fuel requirements of a
process, each kind of fuel can be expressed separately as x
Btu's of coal, y Btu's of oil and gas, and z Btu's of electricity. Data in this disaggregated form is useful in
itself for identifying fuel substitution opportunities within
a process. For comparing fuel consumption impacts of processes or policies, total energy requirements can also be
obtained. In that case, the total fuel requirement is generally expressed as the amount of "primary" energy at the
wellhead (oil and gas) or mine-mouth (coal). In short, the
energy content of fuels, materials, and services is measured
as the energy content of the primary resource embodied in
each and this is a first law measure.

The system boundary which determines which materials and
services are counted (as energy costs of production or energy
impacts of policies) is identical to the boundary which
defines GNP. As Bullard emphasizes, such a system boundary
is anthropocentric and recognizes consumer decisions as the
driving force of production.

To date, the second law measure presented by Ross has
been applied only to the evaluation of direct fuel consumption by technologies, not to the fuel embodied in goods and
services. That measure recognizes that the work which is
potentially available from energy sources is sometimes quite
different from the energy content of the energy source.* In
other words, the amount of work an energy resource can deliver is sometimes quite different than the energy content
would indicate. Second law efficiency is defined as the
ratio of available work consumed by an ideal thermodynamic
system to perform a specified task to the available work
consumed by the actual system used for the task. For example,
a first law efficiency examines how well a natural gas
heater performs in view of its gas consumption. A second
law efficiency examines how well a natural gas heater performs in light of how best to heat a house. The first law
efficiency of natural gas heaters in the U.S. averages 60%
(60% of the energy content of the gas is delivered as heat);
the second law efficiency averages 5% (natural gas heaters

---

*It differs by the Carnot Ratio when the energy is used in
thermal systems.

deliver 5% as much heat as could be delivered by the best*
house heating system).

Note that this performance index depends to a great
extent on just how one defines the task; that is, on where
the boundaries are drawn. Thus, in this example, rather than
asking what is the most second law efficient system for heating a house, one might ask what is the most second law
efficient system for providing all of the energy needs of a
house (hot water, heat, and electricity). While the answer
to the first might be a heat pump, the answer to the second
might well be cogeneration of steam and electricity in
decentralized power plants.

Second law measures relate existing technological
systems to ideal ones via an evaluation of available work
rather than energy content. To date, they have been used
only to evaluate direct fuel consumption.

Concentrated Fuels, Labor, and Environmental Energy

Odum's thesis is that since energy flows are associated
with concentrated fuels, labor, and environmental processes;
in theory, there ought to be a mechanism for evaluating them
and valuing them in energy units. Only with such a comprehensive value measure can we analyze total system energetics,
of which concentrated fuels are only a part. Odum considers
energy analysis to be a part of the basic science of the
energetics of open systems. As such the techniques apply
to environmental systems, technological systems, and economic
systems--to ecosystems, industrial processes, consumer
technologies, cities, regions, nations, or the biosphere as
a whole.

Economics attempts to value concentrated fuels and labor
in money units. Via dollar values, labor substitutions for
fuels and vice versa are made. To date, economics has been
rather ineffective in valuing environmental processes.
Odum's thesis is that energetics can do an effective job of
valuing fuels, labor, and environmental processes and that
substitutions among all three are possible. This section
represents a summary of the theory and techniques developed
by Odum which may make such an evaluation possible.

We have already identified some of the theoretical and
practical difficulties encountered in converting one concentrated fuel type to another using the energy unit. Here
we are faced with the problem of using the energy unit to
convert among fuels, labor, and environmental processes, a
considerably more difficult problem and one that, if it can

---

*Best being defined as thermodynamically ideal.

be resolved, apparently requires some new theory. That
theory, as it is expressed by the maximum power principle,
addresses the empirical question of why systems of any type
or size organize themselves into the patterns observed. Such
a question assumes that physical laws govern system function.
It does not assume, for example, that the system comprising
economic production is driven by consumers; rather that the
whole cycle of production-consumption is structured and
driven by physical laws.

In order to summarize the maximum power principle; one
additional physical concept that is not new must be introduced --the concept of second law efficiency under maximum
power (not to be confused with the second law efficiency
definition given earlier or with the maximum power principle).

Neither the first or second law of thermodynamics include a measure of the rate at which energy transformations
or processes occur. The concept of maximum power incorpo -
rates time into measures of energy transformations. It
provides information about the rate at which one kind of
energy is transformed into another as well as the efficiency
of that transformation. Processing which is 100% efficient
under the second law proceeds at an infinitely slow rate and
processing which proceeds at infinitely fast rates is 0%
efficient. No useful energy is transformed in either case.
If a process is operating under maximum power, it is
delivering the maximum amount of energy possible over a given
amount of time. The efficiency of a process operating at
that rate will always be less than the efficiency of the
ideal second law system.

The theory underpinning energy analysis as developed by
Odum adds another law to these well established ones, one
that is not as yet validated by the vast amount of empirical
evidence supporting the laws of thermodynamics and the concept of second law efficiency under maximum power.

The law has become known as the maximum power principle
and states that systems which maximize their flow of energy
survive in competition. In other words, rather than merely
accepting the fact that more energy per unit of time is
transformed in a process which operates at maximum power,
this principle says that systems organize and structure themselves naturally to maximize power. Systems regulate
themselves according to the maximum power principle. Over
time, the systems which maximize power are selected for
whereas those that do not are selected against and eventually
eliminated.

While not accepted by all ecologists, there is a growing
body of empirical evidence to support such a notion in
ecological systems. Odum argues in his paper that the free
market mechanisms of the economy effectively do the same
thing for human systems and that our economic evolution to

date is a product of that selection process. Some empirical
evidence has accumulated to support that argument.

Consider the oil drilling, production, and refining
system. By this approach, the energy cost of sustaining the
oil production system would be all the energy associated with
fuel usage (both direct and as embodied in materials), with
labor requirements, and with its environmental energy support. Fuel usage is evaluated in a conventional manner where
fuels of different types are converted to energy of equivalent quality (e.g. electricity to its coal equivalent). The
energy cost of the labor is the number of workers multiplied
by their average per capita energy consumption. This assumes
that labor energy consumption is necessary for labor productivity. So, for example, new home appliances and multi cars
per family allow more laborers to work and to be more productive in the economy. That marginal increase in labor and
in labor productivity more than offsets the increased per
capita energy consumption which makes it possible. If it did
not, the new home appliances and multi car phenomena would
not have made it in the market place to begin with.

Environmental energy enters the oil production system
in a diffuse manner throughout the economy. The oil production system requires environmental processes, but the
steel companies which manufacture the steel for the refinery
also require it, and the coal mining industry which mines the
coal to make the coke for steel manufacturing also requires
it. Throughout the economy, processing depends on water
supplied by the hydrological cycle, wind from atmospheric
circulation, oxygen from photosynthesis and so on. Odum's
research on the energy costs of these natural processes
indicates that the contribution of environmental processes
to economic activity averages one-third as large as the
contribution of concentrated fuels. He estimates environmental energies in this manner, giving detailed attention to
some technological systems which appear to be more or less
dependent on environmental processes than the average.

The sum of fuels, labor, and environmental energy represent the energy cost* of oil production. Odum's analysis
indicates that this cost is 1 unit of energy for every 6
units produced, a yield ratio of 6 to 1. These 6 units of
excess energy are available to support other economic activity as well as economic growth. Alternatively, some of the
6 units may be coupled to bring dilute sources of energy,
such as the sun, into production. In that case, the oil effectively amplifies the dilute energy, with the result that
the economy increases its power output further. But the oil is

---

*Energy cost being expressed in units of equivalent quality
such as coal equivalents.

the primary source of energy and the dilute source is
secondary. For example, energy analysis of the Odum type
indicates that solar technology could not survive in the
absence of the oil source on which it depends.

This theory and technique allows one to quantify the
dependence of emerging technologies on existing primary
energy sources. Without government programs, some new technologies would survive and others would not. With this
technique it is possible to predict the relative value of a
new technology to the economy, and, thus, under the maximum
power principle, to predict its survival. Furthermore in
times of energy shortages some old, well established technologies may not survive. If energy production declines
(more specifically if the average yield ratio goes below 6 to
1), then the economy will not be able to afford some of its
parts which did consume the 6 excess units. This technique
purports to predict which parts are more effective in using
the energy and thus to predict which are likely to survive.

Odum's research to date has focused on evaluating the
energy costs and value of two categories of technologies:
emerging energy producing technologies and environmental control technologies. Energy analysis of this type where
concentrated fuels, labor, and environmental processes are
included in the accounting, indicate that many emerging
energy technologies and many environmental control technologies are ineffective. Their energy value to the economy does
not warrant their energy cost in comparison to alternatives
which have the same value at less cost and thus are more
effective. For example, this research indicates that nuclear
power generation (LWR) is a poor alternative to other kinds
of power plants; that it evolved only because of the excess
energy available to support it during exponential energy
growth periods, and that, in the absence of governmental programs which keep it functioning, it would now probably be
eliminated. Similarly, many advanced environmental control
technologies such as tertiary sewage treatment require more
energy than alternative control strategies with the same
effect. More specifically control strategies which couple
technological systems to natural systems are as effective at
less energy cost. For example, sewage treatment systems can
be coupled to natural systems by releasing the effluent onto
swamps or forests; the trees in the forest grow faster and
produce more wood; and the energy cost of this system is
about 25 times less than that of a tertiary treatment plant.
Such an approach recognizes that environmental energy can
substitute for labor and concentrated fuels.

Energy analysis of this type suggests that some new
technologies (and some old ones) which are widely believed
as potentially beneficial are, in fact, ineffective. It
suggests that they will not survive and that governmental

programs which support them will simply lengthen the time
until their demise. By using this form of energy analysis,
Odum maintains that effective alternatives to these can be
identified and their effect quantified. Note that, while
the maximum power principle is deterministic, both empirical
evidence and theory indicate that maximum power is reached
collectively through trial and error of human individuals
acting freely.

Policy Applications

The paper by Gilliland discusses a set of generic policy
questions that energy analyses of the various types can
address. Its purpose is to identify the kinds of information
energy analysis provides that are not provided by other kinds
of analyses. Gilliland's paper is more general than those by
Bullard, Ross and Odum which focus only on those policy
questions to which their respective types of energy analysis
have been applied. Williamson's paper comes at the policy
usefulness of energy analysis from a different point of view.
It discusses the guidelines, programs, and procedures used
by ERDA for the conduct of energy analysis, looks at how
ERDA uses energy analysis in conjunction with other factors
for RD + D planning and decision-making, and evaluates the
utility of energy analysis as compared to other analytical
techniques.

In the following, generic applications (taken from
Gilliland's papers) and some actual applications (taken from
each of the other author's papers) are summarized for each
type of energy analysis. The discussion is divided into
three conceptual categories and an organizational category:
first law applications, second law efficiency applications,
maximum power theory applications, and applications within
ERDA.

First Law Applications

There are a multiplicity of impact concerns raised in
connection with policy-making, for example: environmental,
social, political, and economic impacts. An evaluation of
energy impacts using first law measures adds yet another
important dimension. In these times of energy shortage,
policymakers want to know how, where, and when we use energy
and how, where, and when we can improve energy consumption
patterns. Recall that in this approach, energy is defined
to include only the energy content of concentrated fuels as
they are used directly and indirectly. Governmental data
on energy production and consumption are normally given in
this form.

Thus, this sort of energy analysis can identify process

changes that would increase or decrease energy consumption, identify fuel substitution opportunities in order to alter a fuel mix, measure the energy impact of these process changes or fuel substitutions on production rates and overall energy consumption, and compare the energy costs of alternatives for producing the same commodity. First law analyses, then, can evaluate and compare the energy impacts of public policies which affect energy consumption. Comparative and complementary energy and economic analyses can identify policies which are having unintended influences and suggest new policies. As extended by the Energy Research Group at the University of Illinois, the labor impact of policies can also be measured using a similar technique, but in units of jobs not in energy units.

Eight studies were summarized by Bullard. In some cases technological systems for producing the same commodity were compared (e.g. the energy impacts of throwaway versus refillable containers) and in other cases the evaluation began with a specific policy question (e.g. what are the energy and employment impacts of transferring money from the highway trust fund into other government activities). The analyses quantified the energy costs and the jobs produced by each alternative. For example, the analysis of the recycling question indicated that refilling containers consumes one-half to one-third less energy than manufacturing them. Analysis of the highway trust fund question indicated that the transfer of funds from the highway trust fund into almost any other government activity produced a net increase in jobs and a net decrease in energy consumption. These analyses also identified the interest groups or stakeholders in the policy question. The stakeholders used parts of the results in making public statements on the issue and the researchers testified to legislative bodies clarifying overall results. Such an identification of stakeholders suggests that an evaluation of the energy and employment impacts on each stakeholder may be a potential application of energy analysis. Clearly, impacts are discontinuous; energy and employment impacts are costs to some while they benefit others. Quantifying such discontinuities yields additional policymaking information.

Second Law Efficiency Applications

While the focus of first law analysis is on the quantity of fuel used by existing and proposed technologies as well as that implied by alternative policies, the focus of a second law efficiency analysis of the type described by Ross is on how much energy is required by the tasks that must be carried out. As such, the focus of second law efficiency analysis is on the redesign of systems. The tool has policy utility as an index of performance and as an R + D guide.

As an index of performance, the second law efficiency concept is unique in that it links the kind of energy produced (more precisely the available work in the energy) to the end use work requirements of the task, coupling the two as has not been done previously. If such an index were widely invoked in, for example, product evaluation (if not also in product labeling), more effective comparisons among alternative systems would be possible.

As an R + D guide, the second law efficiency measure forcefully challenges technology to consider a long-range program of improvements in fuel consumption. Because it focuses on performing tasks rather than on processes which already exist and because it compares actual performance to ideal performance, it identifies R + D opportunities for the long term. Research and development activities are often directed at improving existing technologies, for example, the efficiency of car engines, or of natural gas heaters, or of electric power generation. In contrast, second law analyses force such R + D questions as what transportation system is best for moving suburbanites around the Federal Urban Driving Cycle and what system is best for providing the hot water, heating, and electricity needs of a house. The very possibility of being able to examine the performance of systems in light of many definitions of the job to be done identifies R + D opportunities.

Maximum Power Theory Applications

The maximum power theory has many applications in basic science, some of which are identified in Odum's paper. But here we are concerned with public policy applications. Four generic categories of potential policy utility are summarized, each includes some examples from Odum's paper.

First, the theory and techniques allow one to evaluate the value of components of economic systems to entire economic systems. For example, the value of a primary energy source to the economy would be measured by its yield ratio. Primary energy sources now in use have average yield ratios of six to one, which means that the source yields six units of energy to the economy to support more growth for each one unit the economy provides to produce the source. Similarly, the value of system components which are internal to the economic process can be evaluated, by comparing the energy cost of the component to its effect on the economy. Many of these internal components can be considered consumers of the excess units of energy produced from the primary sources; thus one is evaluating how the economy uses the six units of excess energy. As consumers, these internal components consume more energy than they produce. Labor is

an obvious example and the agriculture which supports labor
is another. Present agricultural technology requires two to
ten units of energy for the production of one unit of food
energy. But agriculture and the labor it supports are
clearly necessary for economic vitality; the investment they
require is a measure of their value in contributing to
economic vitality. As less concentrated fuels are tapped,
the yield ratios of primary energy sources are likely to
decline; less than six units of energy for each one produced
will be available to support other economic activity. In the
case where yield ratios are declining, some existing technological systems are likely to disappear for lack of energy to
support them and/or the manner in which they are carried out
will change. Energy intensive agriculture is now under close
scrutiny. This energy analysis technique purports to predict
how systems will change in response to declining primary
energy yields.

The second policy application is in predicting the value
of emerging new technologies to the economy. In the case of
a new primary source, its yield would need to be equal to or
greater than six to one in order to be effective and compete
in today's economic system without government subsidy.
Similarly, new agricultural technology such as greenhouses
would need to require less than or equivalent energy investment than present techniques in order to compete without such
things as food price guarantees.

The third policy application arises because fuels, labor,
and environmental energy, when properly evaluated, are viewed
as substitutions for one another. Thus, the technique can
identify new systems which substitute among fuel, labor, and
environmental energy in such a way that their energy value to
the economy is enhanced. Value is enhanced if the new system
costs less (in energy units) than the alternative with the
same effect. Systems which interface the environment with
technology are currently receiving the most attention.
Properly interfaced, ecosystems apparently can perform much
of the work of processing at considerably less energy cost
than high technology.

Finally, the technique can quantify discontinuous
impacts. For example, the value of a mine-mouth coal-fired
power plant is different for Rosebud County, Montana, than
it is for Montana as a whole, which in turn is different
than its value to the Nation. Energy costs are incurred at
one place while energy value accrues somewhere else. The
electricity from the power plant may have high value to the
Nation but when considered in light of its effect on
Rosebud County's economic vitality, it may have negative
value. The ability to quantify such discontinuities has
many potential policy implications, not the least of which
is the possibility that the Federal government might

consider reimbursements to Rosebud County in proportion to
the discontinuity.

Applications Within the Energy Research
and Development Administration

As discussed in Williamson's paper, procedures currently
used at ERDA evaluate concentrated fuels in energy units,
including direct use of fuels and indirect use of fuels as
they are embodied in materials. Those procedures do not
include an evaluation of labor and environmental energy. The
system boundary for the analysis is the same as the boundary
which defines GNP. Such a boundary excludes much of the
infrastructure of the economy from the accounting. For
example, the energy which supports the Nuclear Regulatory
Commission would not be counted as an energy cost of nuclear
power, using ERDA's procedures. These procedures are similar
to those described by Bullard.

ERDA has supported research which applied these procedures to energy analyses of about 20 energy supply
technologies, 10 end use technologies, and 25 complete energy
trajectories from resource extraction through end use. In
addition, several projects designed to evaluate and improve
energy analysis methodologies have been supported.

ERDA's current focus is on examining and defining the
role of energy analysis in decision processes, particularly
in RD + D decision-making. To date, ERDA has identified
three potential applications of the information provided by
its analyses. First, energy analysis data can help focus
R + D on opportunities for improving energy consumption by
or improving the energy efficiency of energy producing and
consuming technologies. The analyses may indicate both how
much improvement is possible and where in the process R + D
efforts might be most beneficial. Second, energy analysis
data may provide insight into when and at what rate a new,
emerging technology might best be introduced in the market
place. Third, energy analysis data may augment and comple-
ment the results of economic analyses. At this time, ERDA
believes other applications for energy analysis results are
uncertain and indicates that energy analysis is playing a
limited role in R + D decisionmaking. The current ERDA
position is that the superiority of energy analysis to other
analytical techniques has not been proven and that how and
where to include energy analysis in decision processes has
not, as yet, been demonstrated. However, Williamson
indicates that ERDA expects to pursue research both on how
to factor energy analysis into decision processes and on
extensions and refinements of energy analysis methodologies.

Conclusion

Several approaches to energy analysis are now in use. Each is substantially different from the other and each addresses different policy questions and provides different kinds of useful information.

One approach to energy analysis analyzes the energy costs of technologies and the energy impacts of policies by evaluating fuel consumption in terms of the energy content of fuels, goods, and services. The boundary which defines which fuels, goods, and services to count in the analysis is the same boundary which defines GNP. Such a boundary assumes that energy is used directly and embodied in materials as consumers demand it. An extension of the same technique allows labor costs and labor impacts to be analyzed in units of jobs lost and jobs provided. Such analyses provide information about how, where, and when fuels of various types are used in the economy. It provides a means of quantifying the fuel impacts (and labor impacts if the extension is invoked) of alternatives for producing the same commodity and of public policy questions. ERDA is using the results of this kind of analysis and funds research projects of this type.

A second approach analyzes the fuel consumption of technologies in light of the fuel requirements of the task for which the technology is designed. Fuel consumption and fuel requirements are evaluated as available work, a quantity which is sometimes different than the energy content of the fuel. The task definition determines the boundary of the system and only direct fuel consumption is analyzed. Such analyses provide an effective index of performance and suggest R + D opportunities for the redesign of systems better fitted to the requirements of the task.

The third approach recognizes and evaluates the dependence of the economic system as a whole, as well as any part of it, on fuels, labor, and environmental energies. As such, it offers a new theory which might be classed as part of General Systems Theory and as part of the basic science of the energetics of open systems.

The theory suggests that all systems, including our economic system, organize themselves according to physical laws which allow energy output per unit time to be maximized. It suggests that the economic system, rather than being structured by and for consumers, is structured by physical laws for maximum production; individuals provide the creativity which allows maximum power to be attained. The analysis quantifies the dependence of technologies on fuels, labor, and environmental energies (evaluating each in units of energy of equivalent quality) and quantifies the dependence of the economic system on any particular technology or any particular energy source. As such it purports to be able to predict which technological and policy alternatives have the greatest beneficial effect on economic growth and on economic vitality.

The first two approaches have proven useful in addressing an assortment of policy questions having to do with the production and consumption of fuels. The utility of the third approach for addressing value questions as well as the validity of the theory are not widely accepted. The results of many studies which use the third approach challenge present operating norms and have generated controversy and hostility. The question which remains to be answered is, are our conventional models of policy analyses more useful than those being put forward by energy analysts? Presumably, the answer will rest on the power of the competing theories to provide better explanations and therefore better policy information.